MATHÉMATIQUES
&
APPLICATIONS

Directeurs de la collection:
G. Allaire et M. Benaïm

50

T0213442

René Dáger and Enrique Zuazua

Wave Progagation, Observation and Control in 1–*d* Flexible Multi-Structures

 Springer

René Dáger

Departamento de Matemática Aplicada
Universidad Complutense de Madrid
Ciudad Universitaria,s/n
28040 Madrid
Spain
rene_dager@mat.ucm.es

Enrique Zuazua

Departamento de Matemáticas
Facultad de Ciencias, C-XV
Universidad Autónoma de Madrid
Cantoblanco
28049 Madrid
Spain
enrique.zuazua@uam.es

Library of Congress Control Number: 2005930233

Mathematics Subject Classification (2000): 93B05, 93B07, 35L05, 74H45, 93C20

ISSN 1154-483X
ISBN-10 3-540-27239-9 Springer Berlin Heidelberg New York
ISBN-13 978-3-540-27239-9 Springer Berlin Heidelberg New York

Springer est membre du Springer Science+Business Media
© Springer-Verlag Berlin Heidelberg 2006
springeronline.com
Imprimé en Pays-Bas

Imprimé sur papier non acide   41/SPI - 5 4 3 2 1 0 -

Preface

This book is devoted to analyze the vibrations of simplified $1 - d$ models of multi-body structures consisting of a finite number of flexible strings distributed along planar graphs.

We first discuss issues on existence and uniqueness of solutions that can be solved by standard methods (energy arguments, semigroup theory, separation of variables, transposition,...). Then we analyze how solutions propagate along the graph as the time evolves, addressing the problem of the observation of waves. Roughly, the question of observability can be formulated as follows: Can we obtain complete information on the vibrations by making measurements in one single extreme of the network? This formulation is relevant both in the context of control and inverse problems.

Using the Fourier development of solutions and techniques of Nonharmonic Fourier Analysis, we give spectral conditions that guarantee the observability property to hold in any time larger than twice the total length of the network in a suitable Hilbert space that can be characterized in terms of Fourier series by means of properly chosen weights. When the network graph is a tree, we characterize these weights in terms of the eigenvalues of the corresponding elliptic problem. The resulting weighted observability inequality allows identifying the observable energy in Sobolev terms in some particular cases. That is the case, for instance, when the network is star-shaped and the ratios of the lengths of its strings are algebraic irrational numbers.

The observation time we obtain, twice the total length of the network, is optimal. We justify the optimality in the case of a star-shaped network consisting of three strings. We construct a solution, which is the composition of waves with small support, that vanishes at the observation point in a time-interval of length smaller than twice the total length of the network.

These observability results allow us also to solve the problem of controllability, namely, that of driving solutions to rest by means of a control acting on one of the external nodes of the network, using the classical equivalence property between observability and controllability. We describe systematically the control theoretical consequences of the observability properties we have obtained, in terms of the approximate, spectral and exact controllability of networks. More precisely, we deduce sufficient conditions on the network so that a certain subspace (a dense one in the energy space) of initial data may be driven to zero in a time equal to twice the total length of the network. This subspace may be identified to be a Sobolev space under appropriate restrictions on the shape of the network and the lengths of the strings entering in it. More generally, this space may be identified by means of the Fourier series development of solutions on the basis of the eigenfunctions of the Dirichlet laplacian on the network.

The techniques developed to handle this problem and the results we obtain, allow us solving also other similar questions. In particular, the simultaneous observability problem for strings or membranes from an interior region and the control of a network from all its nodes using a small number of different control functions are studied.

Besides, we consider other models on planar networks like Schrödinger, heat or beam-type equations. Existence and uniqueness of solutions is proved in a standard way. We then address the problem of observation and control from an extreme of the network. In order to solve these problems we use various techniques based on the Fourier representation of solutions allowing to derive properties of solutions of those equations as a consequence of those on the wave equation on the same network.

Designed as an introductory course on control and observation of networks, the book contains also some advanced topics which may be of interest for researchers in this area. The last chapter of the book also includes a list of open problems and topics for future research.

Madrid,   *René Dáger*
March 2005   *Enrique Zuazua*

Acknowledgements. This work has been partially supported by Grant BFM2002-03345 of the Spanish MCyT, and the EU TMR Project "Smart Systems".

Contents

1

Introduction

In last years a considerable effort has been devoted to the mathematical study of mechanical systems constituted by coupled flexible or elastic elements as strings, beams, membranes or plates. These systems are known as multi-link or multi-body structures. Their practical relevance is huge. However, the mathematical models describing their evolution are generally quite complex. They can be viewed as systems of Partial Differential Equations (PDE) on networks or graphs.

There is an extensive literature on this topic but a lot remains to be done in order to have a complete theory. Indeed, the interaction between the different components of a multi-link structure may generate new, unexpected pehenomena. Consequently, one can not develop a full theory by simply superposing the existintg results for PDE on domains of the euclidean space. This is particularly true for what concerns control problems. The interested reader is refered to the books [91] and [5] for an introduction the theory of Partial Differential Equations on networks which is an active subject since the early 80's ([82], [83], [97]). In [63] and [68] wide information may be found on modelling and control issues. We also refer to [66] for a systematic analysis of the application of domain decomposition techniques for networks.

But, in view of the intrinsic difficulty of these models it is hard to guess what a general theory should be. It is therefore convenient to first study simplified versions of those models to later address more complex and realistic situations.

This monograph is mainly devoted to analyze the vibrations of a simplified $1-d$ model of a multi-body structure consisting of a finite number of flexible strings distributed along a planar graph. Deformations are assumed to be perpendicular to the reference plane. Though this is an extremely simple and particular model, as we shall see, the whole mathematical picture is quite complex and requires the combination and development of different techniques. We expect the analysis we perform will contribute to clarify what the relevant aspects of the problem are, and to provide some tools for the study of more complex models.

The main goal of this book is to present in a self-contained way the state of the art of the problem of propagation, observation and control of waves on these planar $1 - d$ networks. As we shall see, this requires important developments related with non-harmonic Fourier series, Diophantine approximation, graph theory and wave propagation techniques.

Though the model under consideration is, to some extent, the simplest one in the context of multi-body or multi-link continuous structures, a fine analysis of the nature of the possible vibrations of these planar networks of flexible strings is far from trivial.

The main tool for analyzing the propagation of waves along the graph will be the classical d'Alembert formula, which allows solving the $1 - d$ wave equation both in the space and time directions. In the model under consideration the wave equation holds along each of the strings of the network. The d'Alembert formula allows then representing the solutions on each string explicitly. However, the overall dynamics turns out to be rather complex. This is due to the interaction of the various strings at the junction points. How the energy of waves is transferred from one string to another turns out to be a global problem in which several ingredients arise:

 – the lengths of the various strings constituting the graph;
 – the topology of the graph;
 – the boundary conditions imposed at the extremes of the graph.

The problem of observation or observability concerns, roughly speaking, the issue of determining whether one can determine the total energy of vibrations by partial measurements made for instance, in one or several interior or external nodes of the network. In other words, the property of observability is related with the distribution or propagation of vibrations along the various components of the multi-structure. This problem is relevant, not only because it is a way of analyzing deeply the nature of vibrations, but because it is also of immediate application in the context of inverse and control problems. Part of the book is also devoted to present systematically the consequences of our analysis in what concerns control problems. In particular, we shall analyze the properties of approximate, spectral and exact controllability of networks.

As we mentioned above, graph theory and Diophantine approximation issues enter in a crucial way on the analysis of the property of observability and the topology of the graph plays a fundamental role. For instance, when the graph contains closed circuits there may exist vibrations of the network that remain concentrated and trapped in that circuit, without being propagated to the rest of the network. In those cases, obviously, it is impossible to achieve the observation and/or control property if the observer or controller is not located on the circuit where the solution is trapped. But whether a circuit may support a localized vibration depends also strongly on the mutual lengths of the strings composing the circuit. When all the ratios of the lengths of these strings are rational numbers, such a localized vibration exists. However, if some of these ratios are irrational, then, necessarily, part of the energy of the vibration will be transferred to some other components of the network.

But, in order to determine the amount of energy that is actually transferred one needs to know further properties of that irrational ratio (whether it is algebraic or not, a Liouville number....) and then to apply the existing results on Diophantine approximation.

As we shall see, the overall picture is quite complex, but we hope that this monograph will succeed on describing the main phenomena one may encounter. We shall mainly focus on three cases with different degrees of complexity and such that the corresponding results are also of quite different nature:

The star. It concerns the case where a finite number of strings are connected on a single point by one of their extremes. In this case, using d'Alembert formula, one can give sharp results characterizing the space of observation and/or control in Fourier series by means of suitable weights depending on the lengths of the strings entering in the star-shaped network. We mainly discuss the most difficult case in which observation and/or control are localized in a single extreme of the network. The weights in the corresponding norms depend on the ratios of the lengths of the strings and, in particular, on its irrationality properties. The time needed for observation turns out to be simply twice the sum of all lengths of the strings of the networks.

The tree. It is well known that when all but one external node of the network are observed in a tree-like configuration, the whole energy of solutions may be observed (see [68]). This can be easily seen by an energy argument. Indeed, using sidewise energy estimates for the solutions of the wave equation, one can show that the observation inequality holds in the sharp energy space in a time which is twice the length of the longest path joining the points of the network with some of the observed ends. In this case, the observation time is much smaller than twice the total length of the network, which is needed for the observation from a single end in the case of stars.

Here we analyze the opposite case in which the observation is made at one single extreme of the tree-like network. The observation time turns out to be again, as in the case of one star, twice the sum of the lengths of the strings forming the network. At this point, it is important to note that the case of a tree is a generalization of the previous case of a star. Thus, for the observability property to hold one has also to generalize the condition on the irrationality of the ratios of the lengths of the strings arising in the case of the stars. To do that it is important to observe that the fact of two strings having mutually irrational lengths can also be interpreted in spectral terms. Indeed, it means that the spectra of the two strings have empty intersection. The latter condition turns out to be the appropriate one to be extended to general trees. In this way, the tree turns out to be observable from one end if and only if the spectra of all pairs of subtrees of the tree that match on a nodal point are disjoint. Obviously, this property is also related to the values of the lengths of the strings composing the tree, but does not have an easy interpretation as in the case of the star. Nevertheless, as we shall see, generically, trees satisfy this property.

General networks. The propagation techniques we have employed in the analysis of stars and trees are hard to apply in the case of a network supported by a general graph. Indeed, in the general case we lack of a natural ordering on the graph to analyze the propagation of waves. Actually, as we mentioned above, the presence of closed circuits may trap the waves. Thus, we proceed in a different way by applying a consequence of the celebrated Beurling-Malliavin's Theorem on the completeness of families of real exponentials obtained by Haraux and Jaffard in [50] when analyzing the control of plates. Using the min-max principle, one can show that the spectral density of a general graph is the same as that of a single string whose length is the sum of the lengths of all the strings entering in the network. Then, when the time is greater than twice the total length, as a consequence of Beurling-Malliavin's Theorem, we deduce that there exist some Fourier weights so that the observation property holds in the corresponding weighted norm if and only if all the eigenfunctions of the network are observable. So far we do not know of any necessary and sufficient condition guaranteeing that all the eigenfunctions are observable in the general case. However, this condition, in the particular case of stars and trees discussed above turns out to be sharp: the lengths of the strings are mutually irrational in the case of stars or the spectra of all pairs of subtrees with a common end-point are mutually disjoint in the more general case of trees.

In view of this last result on general networks, the material in this monograph could have been presented in a completely different order. Indeed, we could have started from the most general results on the case of general networks using Beurling-Malliavin's Theorem to later discuss the particular cases of stars and trees using d'Alembert formula and Diophantine approximation, in which general results can be more easily interpreted. However, we have preferred to do all the way around. This corresponds actually to the order and chronology in which the progress was done in the field, starting from the work [75] on the case of a star composed of three strings and continuing with the series of Notes [34, 35, 36, 37].

We became interested on this subject along several discussions with Günter Leugering on this subject and his book in collaboration with Lagnese and Schmidt [68], together with the previously quoted references on PDE on networks, were a great help to start. As we said before, the model we consider in this monograph is the simplest one in the context of vibrations of networks. The interested reader is referred to [68] where many other models can be found with a description of the state of the art in what concerns the well-posedness of the initial boundary problems and the observation and/or control problems for networks of strings, beams, membranes and plates.

Before getting into the analysis of the star we discuss a simpler issue that, nevertheless, allows presenting some of the main difficulties of the theory. It concerns the simultaneous control of two strings connected at one end-point (which is in fact completely equivalent to the problem of controlling one single string from one interior point). In this case we already see the necessity that both strings have mutually irrational lengths. Moreover, we also see that the time needed to control the strings is twice the sum of the lengths of both strings for the observability property to hold. This seems to contradict a first intuition that would suggest that the time needed to control both strings simultaneously should be twice the maximum of the lengths of the strings, i.e., $2\max(\ell_1, \ell_2)$, instead of $2(\ell_1 + \ell_2)$. But, in fact, the time $2(\ell_1 + \ell_2)$ turns out to be sharp under the assumption that the ratio ℓ_1/ℓ_2 is irrational. In other words, even when ℓ_1/ℓ_2 is irrational, the time needed to control simultaneously the two strings together by means of the same control is $2(\ell_1 + \ell_2)$, which is strictly greater than the time needed to control each string independently with two different controls that would be $2\max(\ell_1, \ell_2)$.

It is interesting to analyze the relation of this result with the so-called Geometric Control Condition (GCC) introduced by Bardos, Lebeau and Rauch [18] in the context of the boundary observation and/or control of the wave equation in bounded domains of $\mathbb{R}^n$. The GCC requires that all the rays of Geometric Optics enter the observation region in a finite, uniform time, which turns out to be the minimal one for observation/control. In the case of two strings observed from one common end or the equivalent problem of the string controlled at an interior point, in view of GCC, one could expect the sharp time needed for observation/control to be $2\max(\ell_1, \ell_2)$. But this is not the case, the fact that the rays pass once by the point of observation does not guarantee that the energy concentrated on that ray will be conveniently observed[1]. In fact, we need the ray to pass once more through the point of observation to be able to make a full measurement of the solution. This yields the control/observation time $2(\ell_1 + \ell_2)$. But, in fact, passing twice by the observation point is not sufficient either. The irrationality of the ratio ℓ_1/ℓ_2 is needed to guarantee that, when passing through the observation point the second time, the solution is not exactly at the configuration as in the first crossing, which, of course, would make the second observation to be insufficient too. Finally, even when ℓ_1/ℓ_2 is irrational, we cannot get a uniform bound of the energy of the solution but rather a weaker measurement in a weaker norm. The nature of this norm, which is represented in Fourier series by means of some weights depending on ℓ_1/ℓ_2, depends very strongly on the irrationality class to which the number ℓ_1/ℓ_2 belongs. In fact, in the most favourable case, i.e., when ℓ_1/ℓ_2 is an algebraic number of degree two, one looses one derivative of the solution which, in Sobolev terms means that, for instance, an H^1 observation in time yields only control of the L^2-norm of

---

[1] The wave equation is a second order problem and therefore, even in $1-d$, for a pointwise observation mechanism to be efficient we need to measure not only the position, but also the space derivative. This implies that a necessary condition for observation/control is that all waves pass twice through the observation point.

the solution. In other more pathological cases, like when ℓ_1/ℓ_2 is a Liouville number, one may loose an infinite number of derivatives in the sense that the weights entering in the Fourier representation of the observed norm may have an exponential decay at high frequencies.

We have so far described the content of the main body of the monograph: the propagation, observation and control of waves on stars, trees and general planar networks. But these are only a few of the problems arising in this context. We have complemented this material with the discussion of two important closely related problems:

– The simultaneous observation/control of two strings from a common subinterval. In this case one obtains better results than in the case when the observer/controller is located at a single point [2]. Indeed, this time the results do hold in the sharp energy space without any loss of derivatives. This fact confirms that controlling on an open subinterval is a much more robust mechanism than controlling at a single point.

– The observation/control of general networks through all the nodal points. This is a problem of relevance in applications. From a technological point of view, putting observers/controllers at all the nodal points is feasible. However, one would like to know, for instance, if the number of applied control forces may be reduced by identifying a priori the nodes on which the same force may be applied. This is necessary in order to diminish the complexity of the applied control mechanism. Thus, we would like to know how many different control forces are needed to control the whole structure and to identify the nodes on which each control should be applied. We shall see that the total number of controls needed is four and this is a consequence of our previous analysis and the celebrated Four Colors Theorem.

So far, we have only discussed the wave equation on planar networks of strings. But of course, the same issues arise for all other models like beams, Schrödinger or heat equations. The theory of observation and control of Partial Differential Equations in open domains of $\mathbb{R}^n$ is by now quite well developed (we refer to the survey articles [121] and [123] for an updated account of the developments in this field). However, very little is known in the context of PDE's on networks.

The last part of this monograph is devoted to discuss those three models. Roughly speaking, we show that the results proved in the previous sections on the wave equation yield similar results for those three models. To do that we employ two different results. In the case of the heat equation on the network, we use a classical result by Russell [105] guaranteeing that, whenever the wave equation is controllable in some time, then the heat equation is controllable in an arbitrarily small time. The results of this monograph on the observation and/or control of the wave equation on the network then immediately imply similar results on the corresponding heat model. In what concerns

---

[2] According to the analysis in [42] the problem of pointwise control may be viewed as a singular limit of that of controlling in a subinterval shrinking to that point.

the Schödinger and beams models we use the fact that the time frequencies of the complex exponentials involved in the Fourier representation of solutions of these two models are the squares of those entering in the solutions of the wave equation. Thus, the gap between consecutive eigenfrequencies increases. This allows obtaining observability inequalities for Schrödinger and beam equations from the Fourier representation of those previously obtained for the wave equation. But, this time, as expected, due to the infinite speed of propagation, the observability inequalities hold in an arbitrarily small time.

As we have already mentioned this monograph collects the existing results on simple $1 - d$ models on networks. Much remains to be done in this field. At the end of this book we include a list of open problems and possible subjects of future research. We hope this book to attract the attention to this challenging field of research.

For those who will address these topics for the first time, we refer to [84] for an introduction to some of the most elementary tools on the controllability of PDE's and to the survey articles [121] and [123], for a description of the state of the art in this field.

Finally, some comments on the notations used along this book are in order. The numbering of objects is made locally in each chapter. The sections, subsections, theorems, lemmas, formulas, etc., have a first number to indicate the chapter in which they appear. Thus, Proposition 3.4, is the fourth proposition of Chapter 3. Concerning the constants, they all have been denoted by C. Thus, C may stand for numbers that are different from line to line of the text, but that remain uniform with respect to the relevant parameters. Only when we intend to explicitly indicate the dependence of C on some parameter, or to avoid ambiguities, we use some more complete notations.

We would like to emphasize that the book is mainly self-contained and that it has been designed as an introductory course to the controllability and observability of networks for graduate students. The text may be covered in the order presented or, if a simplified approach is desired, it is possible to restrict oneself to Chapters 2, 3, 4 and 8, as the remaining chapters are more technical. However, many other variants are also possible.

2

Preliminaries

2.1 The Elastic String

Let us start with a simple example. Consider an elastic string of length one which is fixed at its ends. The deformation of the string is given by the function $\phi(t, x) : \mathbb{R} \times (0, 1) \to \mathbb{R}$ which is the unique solution of the wave equation

$$
\begin{aligned}
\phi_{tt} - \phi_{xx} &= 0 && \text{in } \mathbb{R} \times (0, 1), \\
\phi(t, 0) = \phi(t, 1) &= 0 && \text{in } \mathbb{R}, \\
\phi(0, x) = \phi_0(x), \quad \phi_x(0, x) &= \phi_1(x) && \text{in } (0, 1),
\end{aligned}
\tag{2.1}
$$

where ϕ_0 and ϕ_1 are the initial deformation and velocity of the string, respectively.

The solution of system (2.1) may be expressed by the Fourier formula

$$
\phi(t, x) = \sum_{n=1}^{\infty} (a_n \cos n\pi t + \frac{b_n}{n\pi} \sin n\pi t) \sin n\pi x,
\tag{2.2}
$$

where (a_n) and (b_n) are the sequences of Fourier coefficients in the orthogonal basis of $L^2(0, 1)$:

$$
\theta_n(x) = \sin n\pi x, \quad n = 1, 2, \ldots.
$$

The energy of the solution ϕ is defined as

$$
E_\phi(\phi_0, \phi_1, t) = \frac{1}{2} \int_0^1 \left(|\phi_x(t, x)|^2 + |\phi_t(t, x)|^2 \right).
$$

It is easy to prove that the energy of a solution is constant[1], that is $E_\phi(t) = E_\phi(0)$. The energy is a norm in the space $H_0^1(0, 1) \times L^2(0, 1)$ of initial states

---

[1] This can be done computing directly on the Fourier representation of the solution or, by the energy method, i.e. multiplying the wave equation by ϕ_t and integrating with repesct to x. After integration by parts this yields $dE(t)/dt = 0$.

of (2.1) and may be expressed in terms of the Fourier coefficients (a_n) and (b_n) as

$$E_\phi(\phi_0, \phi_1) = \frac{1}{4} \sum_{n=1}^{\infty} (n^2 \pi^2 a_n^2 + b_n^2). \tag{2.3}$$

Assume now that we observe the motion of the string at one of its points. To fix ideas, suppose we know the values of the velocity ϕ_t and the tension ϕ_x at some point $x = \xi$ in a time interval $(0, T)$. Let us define the *observation function*

$$\Phi(\phi_0, \phi_1, \xi, T) = \frac{1}{4} \int_0^T |\phi_t(t, \xi)|^2 dt + \frac{1}{4} \int_0^T |\phi_x(t, \xi)|^2 dt.$$

Let us note that for $T = 2M$ with $M \in \mathbb{N}$ it holds

$$\Phi(\phi_0, \phi_1, \xi, T) = M E_\phi(\phi_0, \phi_1). \tag{2.4}$$

Indeed, from the formula (2.2) we have

$$\phi_t(t, \xi) = \sum_{n=1}^{\infty} (-n\pi a_n \sin n\pi t + b_n \cos n\pi t) \sin n\pi \xi,$$

$$\phi_x(t, \xi) = \sum_{n=1}^{\infty} (n\pi a_n \cos n\pi t + b_n \sin n\pi t) \cos n\pi \xi$$

and then, in view of the 2-periodicity of the functions $\sin n\pi t$ and $\cos n\pi t$ and their orthogonality properties,

$$\int_0^{2M} |\phi_t(t, \xi)|^2 dt = M \int_0^2 |\phi_t(t, \xi)|^2 dt = M \sum_{n=1}^{\infty} (n^2 \pi^2 a_n^2 + b_n^2) \sin^2 n\pi \xi, \tag{2.5}$$

$$\int_0^{2M} |\phi_x(t, \xi)|^2 dt = M \int_0^2 |\phi_x(t, \xi)|^2 dt = M \sum_{n=1}^{\infty} (n^2 \pi^2 a_n^2 + b_n^2) \cos^2 n\pi \xi. \tag{2.6}$$

Therefore, in view of (2.3), (2.5) and (2.6) we obtain (2.4).

Clearly, the function $\Phi(\phi_0, \phi_1, \xi, T)$ is increasing in T, so, if $2 \leq T \leq 2M$ with $M \in \mathbb{N}$ we obtain

$$\Phi(\phi_0, \phi_1, \xi, 2) \leq \Phi(\phi_0, \phi_1, \xi, T) \leq \Phi(\phi_0, \phi_1, \xi, 2M),$$

or equivalently,

$$E_\phi(\phi_0, \phi_1) \leq \Phi(\phi_0, \phi_1, \xi, T) \leq M E_\phi(\phi_0, \phi_1).$$

That means that, for all $\xi \in [0, 1]$ and $T \geq 2$, the norms defined by E_ϕ and $\Phi(\cdot, \xi, T)$ are equivalent. That is, it is possible to estimate the energy of the solution ϕ from the measurements of ϕ_t, ϕ_x made at point ξ during a time interval of length at least two. In particular, when $T = 2$ those two norms coincide:

$$E_\phi(\phi_0, \phi_1) = \Phi(\phi_0, \phi_1, \xi, 2).$$

When $\xi = 0$ or $\xi = 1$, the observation function Φ is simpler. For instance, for $x = 0$ it becomes

$$\Phi(\phi_0, \phi_1, 0, T) = \frac{1}{4} \int_0^T |\phi_x(t, 0)|^2 dt,$$

since $\phi_t(t, 0) \equiv 0$.

Accordingly, at the boundary points, the observation of the tension ϕ_x of the string during a time-interval of length twice the length of the string, suffices to fully recover the total energy of the vibration.

It is natural to raise the question of whether the same happens at the internal observation points ξ. Accordingly, consider a weaker observation function:

$$\Psi(\phi_0, \phi_1, \xi, T) = \frac{1}{4} \int_0^T |\phi_x(t, \xi)|^2 dt.$$

We already know that, when $\xi = 0$ or $\xi = 1$ this function defines a norm in the space of initial data, equivalent to the energy-norm. The following questions arise naturally: *does the function Ψ define a norm in $H_0^1(0,1) \times L^2(0,1)$? If so, is that norm equivalent to the energy?*

Assume $T = 2$, then in view of (2.6) it holds

$$\Psi(\phi_0, \phi_1, \xi, 2) = \frac{1}{4} \sum_{n=1}^\infty (n^2 \pi^2 a_n^2 + b_n^2) \cos^2 n\pi\xi. \tag{2.7}$$

Formula (2.7) is very similar to (2.3) and, clearly,

$$\sum_{n=1}^\infty (n^2\pi^2 a_n^2 + b_n^2) \cos^2 n\pi\xi \leq \sum_{n=1}^\infty (n^2\pi^2 a_n^2 + b_n^2),$$

and then

$$\Psi(\phi_0, \phi_1, \xi, 2) \leq E_\phi(\phi_0, \phi_1).$$

However, the converse inequality is not true whatever $\xi \in (0,1)$ is. Indeed, the converse inequality would require a lower bound of the form

$$|\cos n\pi\xi| \geq C, \tag{2.8}$$

for every $n \in \mathbb{N}$. But this inequality is false for all $\xi \in (0,1)$. Indeed, if ξ is a rational number that can be expressed as

$$\xi = \frac{2p+1}{2q}, \qquad p, q \in \mathbb{Z}, \tag{2.9}$$

then, when $n = qk$ with k odd

$$\cos n\pi\xi = \cos \frac{(2p+1)\,k}{2}\pi = 0.$$

Thus, in this case, $\cos n\pi\xi = 0$ for an infinite number of values of n and consequently, inequality (2.8) cannot be true. That means that the function $\Psi(\cdot, \xi, 2)$ is not even a norm in $H_0^1(0,1) \times L^2(0,1)$.

On the other hand, when the number ξ cannot be expressed in the form (2.9) all the numbers $\cos n\pi\xi$ are different from zero. This implies that the function $\Psi(\cdot, \xi, 2)$ does define a norm in $H_0^1(0,1) \times L^2(0,1)$. But this norm is necessarily weaker than the energy.

In fact, inequality (2.8) is equivalent to the existence of a positive number α such that, for all $k, n \in \mathbb{Z}$,

$$\left| n\pi\xi - \frac{2k+1}{2}\pi \right| \geq \alpha.$$

That is

$$|(2\xi)\,n - (2k+1)| \geq \alpha_0 := \frac{2\alpha}{\pi}.$$

This rational approximation property of the number 2ξ is false for all $\xi \in (0,1)$. We will discuss this issue in detail in Chapter 3.

But, for certain values of ξ weaker inequalities may be obtained. Indeed, for instance, if 2ξ may be expanded in continuous fraction $[0, c_1, c_2, ...]$ with bounded sequence (c_n) then there exists a constant C_ξ such that

$$|(2\xi)\,n - (2k+1)| \geq C_\xi/n,$$

and this is the best lower bound one may expect. This implies that

$$|\cos n\pi\xi| \geq C_\xi/n$$

and therefore

$$\Psi(\phi_0, \phi_1, \xi, 2) \geq C_\xi \sum_{n=1}^{\infty} (a_n^2 + \frac{b_n^2}{n^2\pi^2}) = C_\xi \|\phi_0\|^2_{L^2(0,1)} + \|\phi_1\|^2_{H^{-1}(0,1)}.$$

Summarizing, for the values of ξ indicated above, it holds

$$C_\xi \left(\|\phi_0\|^2_{L^2(0,1)} + \|\phi_1\|^2_{H^{-1}(0,1)} \right) \leq \Psi(\phi_0, \phi_1, \xi, 2) \leq \|\phi_0\|^2_{H_0^1(0,1)} + \|\phi_1\|^2_{L^2(0,1)}.$$

This is the best result we may obtain. Accordingly, for interior points $\xi \in (0,1)$, the information contained in $\Psi(\phi_0, \phi_1, \xi, 2)$ does not suffice to recover the whole energy of the string and only weaker norms may be recovered (the $[L^2(0,1) \times H^{-1}(0,1)]$-nom in the particular case above with a loss of one derivative in $L^2(0,1)$, both for ϕ and ϕ_t). This is also the case when considering other kind of observation functions, e.g.,

$$\int_0^T |\phi(t,\xi)|^2 dt.$$

As we shall see in the following chapters, this is the typical situation when addressing the problem of observability for the vibrations of a network of strings. Typicallly, one can recover only weaker energies from measurements made at some points of the strings, even if at those points both the velocity and the tension are measured[2].

When the observation is made on a larger set, say on some interval $\omega \subset (0,1)$, then the total energy can be recovered. Indeed, consider the observation function

$$\int_0^T \int_\omega |\phi_x(t,x)|^2 dx dt.$$

Assume that $T = 2$. Then

$$\int_0^2 \int_\omega |\phi_x(t,x)|^2 dx dt = \int_\omega \int_0^2 |\phi_x(t,x)|^2 dt dx$$

$$\geq \sum_{n=1}^\infty (n^2\pi^2 a_n^2 + b_n^2) \int_\omega \sin^2 n\pi x \; dx. \tag{2.10}$$

But, for any $\omega \subset (0,1)$ there exists a constant $C_\omega > 0$ such that

$$\int_\omega \sin^2 n\pi x \; dx \geq C_\omega$$

for every $n \in \mathbb{N}$. Therefore,

$$C_\omega \sum_{n=1}^\infty (n^2\pi^2 a_n^2 + b_n^2) \leq \int_0^2 \int_\omega |\phi_x(t,x)|^2 dx dt \leq |\omega| \sum_{n=1}^\infty (n^2\pi^2 a_n^2 + b_n^2),$$

that is

$$4C_\omega E_\phi \leq \int_0^2 \int_\omega |\phi_x(t,x)|^2 dx dt \leq 4|\omega|E_\phi.$$

Using the d'Alembert formula for the representation of the solutions of the wave equation, one may improve the estimate above on the time needed for this estimate to be true. Namely, the property

$$C_1 E_\phi \leq \int_0^T \int_\omega |\phi_x(t,x)|^2 dx dt \leq C_2 E_\phi,$$

holds for any $T > 2\mathrm{dist}\{\omega, \{0,1\}\}$, for some positive constants C_1 and C_2. The time $2\mathrm{dist}\{\omega, \{0,1\}\}$ is actually the characteristic one and it is in agreement

---

[2] As we shall see, there is a case in which this does not happen and the whole energy may be recovered: For tree-like networks when the tension is measured in all the external nodes except at most one.

with the Geometric Control Condition (GCC) mentioned in the introduction that indicates that, for the observability inequality to hold, all rays should enter the observation region in the given observation time.

But for networks of strings, observing on a subinterval of one of the strings will not help. This allows recovering the information on the string where the observation is being made but will only yield weaker measurements on the other ones.

2.2 Networks of Strings

2.2.1 Elements on Graphs

A graph G is a pair $(\mathcal{V}, \mathcal{E})$, where $\mathcal{V}$ is a set, whose elements are called *vertices* of G, and $\mathcal{E}$ is a family of non-ordered pairs $\mathbf{v}, \mathbf{w}$ of vertices, which we will denote by $\widehat{\mathbf{vw}}$. The elements of $\mathcal{E}$ are called *edges* of G with vertices $\mathbf{v}, \mathbf{w}$. When the graph G does not contain edges of the form $\widehat{\mathbf{vv}}$ it is said that the graph is *simple*[3].

A *path* between the vertices $\mathbf{v}$ and $\mathbf{w}$ of a graph G is a set of edges of the form

$$\widehat{\mathbf{vv}_1}, \widehat{\mathbf{v}_1\mathbf{v}_2}, ..., \widehat{\mathbf{v}_{m-1}\mathbf{v}_m}, \widehat{\mathbf{v}_m\mathbf{w}}.$$

If all the edges forming a path are different, it is said that the path is *simple*; if all the vertices $\mathbf{v}_1, ..., \mathbf{v}_m$ are different, the path is called *elementary*.

A *closed path* is a path between a vertex and itself. An elementary closed path is called a *cycle*. When the graph G does not contain cycles it is said that G is a *tree*.

Graphs with a finite number of vertices are called *finite*. In this book we shall be concerned only with finite graphs.

Let us suppose that G is a finite graph with N vertices and M edges:

$$\mathcal{V} = \{\mathbf{v}_1, ..., \mathbf{v}_N\}, \qquad \mathcal{E} = \{\mathbf{e}_1, ..., \mathbf{e}_M\}.$$

The multiplicity $m(\mathbf{v})$ of the vertex $\mathbf{v}$ is the number of edges that meet at $\mathbf{v}$:

$$m(\mathbf{v}) := \operatorname{card}\{\mathbf{e} \in \mathcal{E}: \quad \mathbf{v} \in \mathbf{e}\}.$$

We also define the sets

$$\mathcal{V}_{\mathcal{S}} := \{\mathbf{v} \in \mathcal{V}: \quad m(\mathbf{v}) = 1\}, \quad \mathcal{V}_{\mathcal{M}} := \mathcal{V} \setminus \mathcal{V}_{\mathcal{S}},$$

where $\mathcal{V}_{\mathcal{S}}$ is the set of those vertices that belong to a single edge, the *exterior* ones, while $\mathcal{V}_{\mathcal{M}}$ contains the remaining vertices, the *interior* ones, i.e., those that belong to more than one edge.

---

[3] Sometimes the term graph is used only for simple graphs, that is, for those that do not have edges with equal vertices. Non-simple graphs are then called *pseudographs*.

For a vertex $\mathbf{v}$ we denote by

$$I_{\mathbf{v}} := \{i : \mathbf{v} \in \mathbf{e}_i\},$$

the set of indices of all those edges of G which are incident to $\mathbf{v}$. If the vertex $\mathbf{v}_j$ is exterior, $I_{\mathbf{v}_j}$ contains a single index; it will be denoted by $i(j)$ and, if this does not lead to misunderstanding, simply by i.

In this book we consider only simple finite graphs whose vertices are points of a plane. The edges of the graph are viewed as rectilinear segments joining some of those points. The length of the segment corresponding to the edge $\mathbf{e}_i$ is called length of $\mathbf{e}_i$ and is denoted by ℓ_i.

We will also assume that the edges of the graphs may meet only at the vertices of G. Such graphs are known as *planar graphs*.

On every edge of G we choose an orientation (that is, one of the vertices has been chosen as the initial one). Then $\mathbf{e}_i$ may be parametrized as a function of its arc length by means of the functions $x_i : [0, \ell_i] \to \mathbf{e}_i$.

We define the incidence matrix of G

$$\varepsilon_{ij} = \begin{cases} \text{-1} & \text{if } x_i(0) = \mathbf{v}_j, \\ +1 & \text{if } x_i(\ell_i) = \mathbf{v}_j. \end{cases}$$

Let us denote by L the sum of the lengths of all the edges of the graphs, the *length of the graph*. To indicate to which graph it corresponds, we shall write, if necessary, L_G.

Given functions $u^i : [0, \ell_i] \to \mathbb{R}$, $i = 1, ..., M$, we will denote by $\bar{u} : G \to \mathbb{R}$ the function defined for $\mathbf{x} \in \mathbf{e}_i$ by

$$\bar{u}(\mathbf{x}) = u^i(x_i^{-1}(\mathbf{x})).$$

In this case, we will say that $\bar{u}$ is a function defined on the graph G with components u^i. Frequently, we will indicate this fact just by writing $\bar{u} = (u^1, ..., u^M)$. In particular, the vector with null components will be denoted by $\bar{0}$.

2.2.2 Equations of Motion for Networks

Now we consider a planar network of elastic strings that undergo small perpendicular vibrations. At rest, the network coincides with a planar graph G contained in that plane.

Let us suppose that the function $u^i = u^i(t, x) : \mathbb{R} \times [0, \ell_i] \to \mathbb{R}$, describes the transversal displacement in time t of the string that coincides at rest with the edge $\mathbf{e}_i$. Then, for every $t \in \mathbb{R}$, the functions u^i, $i = 1, ..., M$, define a function $\bar{u}(t)$ on G with components $u^i : \mathbb{R} \times [0, \ell_i] \to \mathbb{R}$ given by $u^i(x, x) = u^i(t, x_i(x))$. This function allows to identify the network with its rest graph; in this sense, the vertices of G will be called nodes and the vertices, strings.

As a model of the motion of the network we assume that the displacements u^i satisfy the following non-homogeneous system

$$u^i_{tt} - u^i_{xx} = 0 \qquad\qquad\qquad \text{in } \mathbb{R} \times [0, \ell_i], \quad i = 1, ..., M, \qquad (2.11)$$

$$u^{i(j)}(t, \mathbf{v}_j) = h_j(t) \qquad\qquad t \in \mathbb{R}, \quad j = 1, ..., r, \qquad\qquad (2.12)$$

$$u^{i(j)}(t, \mathbf{v}_j) = 0 \qquad\qquad\quad t \in \mathbb{R}, \quad j = r+1, ..., N, \qquad (2.13)$$

$$u^i(t, \mathbf{v}) = u^j(t, \mathbf{v}) \qquad\qquad t \in \mathbb{R}, \quad \mathbf{v} \in \mathcal{V}_{\mathcal{M}}, \ i, j \in I_{\mathbf{v}}, \qquad (2.14)$$

$$\sum\nolimits_{i \in I_{\mathbf{v}}} \partial_n u^i(t, \mathbf{v}) = 0 \qquad\quad t \in \mathbb{R}, \quad \mathbf{v} \in \mathcal{V}_{\mathcal{M}}, \qquad\qquad (2.15)$$

$$u^i(0, x) = u^i_0(x), \ u^i_t(0, x) = u^i_1(x) \quad x \in [0, \ell_i], \quad i = 1, ..., M, \qquad (2.16)$$

where $\mathcal{C} = \{\mathbf{v}_1, ..., \mathbf{v}_r\}$ is a non-empty subset of $\mathcal{V}_{\mathcal{S}}$ (the set of controlled nodes) and $\partial_n u^i(t, \mathbf{v}) := \varepsilon_{ij} u^i_x(t, x_i^{-1}(\mathbf{v}))$ is the exterior normal derivative of u_i at the node $\mathbf{v}$.

Thus, (2.11)-(2.16) corresponds to a network with r controlled exterior nodes. The time-dependent functions $h_j = h_j(t)$ are the controls that act on the system through the vertices $\mathbf{v}_j$, $j = 1, ..., r$.

Equation (2.11) is the classical 1-d wave equation, which is verified by the deformations of the strings of the network. The equalities (2.12), (2.13) reflect the condition that over some of the exterior nodes, precisely over those corresponding to the vertices contained in $\mathcal{C}$, some controls act to regulate their displacements, while the remaining nodes are fixed. The relations (2.14) and (2.15) express the continuity of the network and the balance of forces at the interior nodes. Finally, (2.16) imposes the initial deformation and velocity of the strings (i.e., at time $t = 0$). The pair $(\bar{u}_0, \bar{u}_1)$ is called *initial state* of the network.

In general, we will suppose that the graph G does not contain vertices of multiplicity two, since those would be irrelevant in our model. Indeed, they may be considered as interior points of an edge whose length coincides with the sum of the lengths of the edges coupled at that vertex.

In order to study system (2.11)-(2.16), we need a proper functional setting. We define the Hilbert spaces

$$V = \{\bar{u} \in \prod_{i=1}^{M} H^1(0, \ell_i) : u^i(\mathbf{v}) = u^j(\mathbf{v}) \text{ if } \mathbf{v} \in \mathcal{V}_{\mathcal{M}} \text{ and } u^i(\mathbf{v}) = 0 \text{ if } \mathbf{v} \in \mathcal{V}_{\mathcal{S}}\},$$

$$H = \prod_{i=1}^{M} L^2(0, \ell_i),$$

endowed with the Hilbert structures

$$< \bar{u}, \bar{w} >_V := \sum_{i=1}^{M} < u^i, w^i >_{H^1(0, \ell_i)} = \sum_{i=1}^{M} \int_0^{\ell_i} u^i_x w^i_x dx,$$

$$< \bar{u}, \bar{w} >_H := \sum_{i=1}^{M} < u^i, w^i >_{L^2(0, \ell_i)} = \sum_{i=1}^{M} \int_0^{\ell_i} u^i w^i dx,$$

respectively. Besides, we will denote

$$U = \left(L^2(0,T)\right)^r,$$

the space of controls.

Since the injection $V \subset H$ is dense and compact, when H is identified with its dual H' by means of the Riesz-Fréchet isomorphism, we can define the operator $-\Delta_G : V \to V'$ by

$$\langle -\Delta_G \bar{u}, \bar{v} \rangle_{V' \times V} = \langle \bar{u}, \bar{v} \rangle_H.$$

The operator $-\Delta_G$ is an isometry from V to V'. The notation $-\Delta_G$ is justified by the fact that, for smooth functions $\bar{u} \in V$, the operator $-\Delta_G$ coincides with the Laplace operator.

It may be shown that the spectrum of the operator $-\Delta_G$ is formed by an increasing positive sequence $(\mu_n)_{n \in \mathbb{N}}$ of eigenvalues. The corresponding eigenfunctions $(\bar{\theta}_n)_{n \in \mathbb{N}}$ may be chosen to form an orthonormal basis of H.

The spaces V and H may be characterized as

$$V = \left\{ \bar{u} = \sum_{n \in \mathbb{N}} u_n \bar{\theta}_n : \quad |||\bar{u}|||_V^2 := \sum_{n \in \mathbb{N}} \mu_n u_n^2 < \infty \right\},$$

$$H = \left\{ \bar{u} = \sum_{n \in \mathbb{N}} u_n \bar{\theta}_n : \quad |||\bar{u}|||_H^2 := \sum_{n \in \mathbb{N}} u_n^2 < \infty \right\},$$

and the norms of V and H are equivalent to $|||.|||_V$ and $|||.|||_H$, respectively. The spaces V and H are Hilbert spaces with respect to the scalar products that generate the corresponding norms.

The study of the solvability of system (2.11)-(2.16) may be done in the standard way by the classical *transposition method* (see [80]). Let us describe the main steps, since some of its elements are widely used in through this book. The details of this procedure may be found in [80] or [59]. The application of this technique to the particular problem of string networks may be found in [68].

First we study the homogeneous problem ($h_j \equiv 0$ for all $j = 1, ..., r$)

$$\phi_{tt}^i - \phi_{xx}^i = 0 \qquad \text{in } \mathbb{R} \times [0, \ell_i], \quad i = 1, ..., M, \qquad (2.17)$$

$$\phi^{i(j)}(t, \mathbf{v}_j) = 0 \qquad t \in \mathbb{R}, \quad j = 1, ..., N, \qquad (2.18)$$

$$\phi^i(t, \mathbf{v}) = \phi^j(t, \mathbf{v}) \qquad t \in \mathbb{R}, \quad \mathbf{v} \in \mathcal{V}_\mathcal{M}, \; i, j \in I_{\mathbf{v}}, \qquad (2.19)$$

$$\sum_{i \in I_{\mathbf{v}}} \partial_n \phi^i(t, \mathbf{v}) = 0 \qquad t \in \mathbb{R}, \quad \mathbf{v} \in \mathcal{V}_\mathcal{M}, \qquad (2.20)$$

$$\phi^i(0, x) = \phi_0^i(x), \quad \phi_t^i(0, x) = \phi_1^i(x) \quad x \in [0, \ell_i], \quad i = 1, ..., M. \qquad (2.21)$$

The solution of the homogeneous system (2.17)-(2.21) with initial data

$$\bar{\phi}_0 = \sum_{n \in \mathbb{N}} \phi_{0,n} \bar{\theta}_n, \qquad \bar{\phi}_1 = \sum_{n \in \mathbb{N}} \phi_{1,n} \bar{\theta}_n, \tag{2.22}$$

is then defined by the formula

$$\bar{\phi}(t,x) := \sum_{n \in \mathbb{N}} \left(\phi_{0,n} \cos \lambda_n t + \frac{\phi_{1,n}}{\lambda_n} \sin \lambda_n t \right) \bar{\theta}_n(x). \tag{2.23}$$

For a classical smooth solution $\bar{u}$ of (2.11)-(2.16), the energy is defined as the sum of the energies of its components, that is,

$$\mathbf{E}_{\bar{u}}(t) := \sum_{i=1}^{M} \mathbf{E}_{u_i}(t) \quad \text{with} \quad \mathbf{E}_{u_i}(t) := \frac{1}{2} \int_0^{\ell_i} \left(|u_t^i(t,x)|^2 + |u_x^i(t,x)|^2 \right) dx.$$

From equations (2.11)-(2.15), it is easily proved that

$$\frac{d}{dt} \mathbf{E}_{\bar{u}}(t) = \sum_{i=1}^{r} u_t^i(t,\mathbf{v}_j) \partial_n u^i(t,\mathbf{v}_j). \tag{2.24}$$

In particular, in the homogeneous case (2.17)-(2.21), the energy is conserved:

$$\mathbf{E}_{\bar{\phi}}(t) = \mathbf{E}_{\bar{\phi}}(0),$$

for every $t \in \mathbb{R}$. Besides, if the initial data are as in (2.22) then

$$\mathbf{E}_{\bar{\phi}} = \frac{1}{2} \sum_{n \in \mathbb{N}} (\mu_n \phi_{0,n}^2 + \phi_{1,n}^2) = \frac{1}{2} (|||\bar{\phi}_0|||_V^2 + |||\bar{\phi}_1|||_H^2). \tag{2.25}$$

From the definition (2.23) it holds that, for all $T \in \mathbb{R}$ and $(\bar{\phi}_0, \bar{\phi}_1) \in V \times H$, the solution $\bar{\phi}$ satisfies

$$\bar{\phi} \in C([0,T] : V) \bigcap C^1([0,T] : H). \tag{2.26}$$

In addition, $\bar{\phi}$ is the unique solution of the system (2.17)-(2.21) in the sense of distributions, which has the property (2.26).

In what follows, we denote by S_t the group of isometries generated by (2.17)-(2.21).

For every $s \in \mathbb{R}$ we consider the Hilbert spaces

$$V^s := \left\{ \bar{u} = \sum_{n \in \mathbb{N}} u_n \bar{\theta}_n : \quad \|\bar{u}\|_s^2 := \sum_{n \in \mathbb{N}} \mu_n^s u_n^2 < \infty \right\}, \tag{2.27}$$

$$h^s := \left\{ (u_n) : \quad \|(u_n)\|_s^2 := \sum_{n \in \mathbb{N}} \mu_n^s |u_n|^2 < \infty \right\}, \tag{2.28}$$

endowed with the norms $\|\cdot\|_s$, where (u_n) denotes a sequence of real numbers u_n. The canonical isomorphism $\sum_{n \in \mathbb{N}} u_n \bar{\theta}_n \to (u_n)$ is an isometry between V^s and h^s.

Let us observe that V^s is the domain of $(-\Delta_G)^{\frac{s}{2}}$ considered as an unbounded operator from H to H. Besides, $V = V^1$ and $H = V^0$.

Further, we introduce the Hilbert spaces

$$\mathcal{W}^s := V^s \times V^{s-1},$$

endowed with the natural product structures. We then have

$$\mathcal{W}^1 = V \times H, \qquad \mathcal{W}^0 = H \times V'.$$

For initial state $(\bar{\phi}_0, \bar{\phi}_1) \in \mathcal{W}^s$ the solution of the homogeneous problem (2.17)-(2.21) may be defined by (2.26). In this case,

$$\bar{\phi} \in C([0,T] : V^s) \bigcap C^1([0,T] : V^{s-1}).$$

The following step in the study of the solvability of system (2.11)-(2.16) consists in proving that, for every $T > 0$, there exists a constant $C > 0$ such that, at every exterior node $\mathbf{v} \in \mathcal{V}_S$, the smooth solutions of (2.17)-(2.21) satisfy the inequality

$$\int_0^T |\partial_n \phi^i(t, \mathbf{v})|^2 dt \le C \mathbf{E}_{\bar{\phi}}. \tag{2.29}$$

This property is known as *hidden regularity*, since it is not a consequence of (2.26) and standard trace results; it is an specific property of the solutions of (2.17)-(2.21) and, more generally, of the Dirichlet problem for wave equations. The inequality (2.29) may be proved using D'Alembert formula for the representation of the solutions of (2.17). In [68], this inequality is proved by means of the multiplier technique, which is also useful in the wider context of wave equations in several dimensions and having smooth variable coefficients.

In what follows, in order to simplify the notations, we suppose that $r = 1$, that is, only one node of the network is controlled, the one corresponding to the index $i = 1$. Obviously, this is the most delicate situation for controllability to hold. When the number of controls increases, the controllability properties of the system are enhanced.

Fix $T > 0$ and define for every $t \in (0, T]$ the operator $\mathbf{A}_t : H \times V \to L^2(0, T)$, which associates to every pair $(\bar{\phi}_1, -\bar{\phi}_0) \in H \times V$ the normal derivative $\partial_n \phi^1(., \mathbf{v}_1)$ in the controlled node of the solution (2.23) of the non-homogeneous system (2.17)-(2.21).

In view of (2.29) and (2.25), $\mathbf{A}_t$ is continuous. Then, the operator $\mathbf{A}_t^* : L^2(0, T) \to H \times V'$, the adjoint of $\mathbf{A}_t$, is also continuous (we have identified $L^2(0, T)$ and H with their duals).

Further, for every $h \in L^2[0, T]$ we define the solution of system (2.11)-(2.16) with null initial state $(\bar{u}_0, \bar{u}_1) \in H \times V'$ as

$$\bar{u} = \mathbf{A}_t^* h + \mathbf{S}_t(\bar{u}_0, \bar{u}_1), \tag{2.30}$$

where $\mathbf{S}_t(\bar{u}_0, \bar{u}_1)$ is the solution with null controls, i.e. $h_j \equiv 0$, $j = 1, ..., r$.

To clarify the meaning of this formula, let us calculate the operator $\mathbf{A}_t^*$. We consider the operator $\mathbf{B}$ defined for $h \in C^1([0, t])$ by

$$\mathbf{B}h = (\bar{u}(t), \bar{u}_t(t)),$$

where $\bar{u}$ is the solution in the classical sense of the problem (2.11)-(2.16) with null initial data $\bar{u}_0 = \bar{u}_1 = 0$.

If we multiply equation (2.11) by u_i and integrate over $[0, t] \times [0, \ell_i]$ it holds, after integration by parts,

$$\int_0^t \int_0^{\ell_i} (u_{tt}^i - u_{xx}^i)\phi^i dt dx = - \int_0^{\ell_i} \left(u^i\phi_t^i - u_t^i\phi^i \right)\big|_0^t dx + \int_0^t \left(u_x^i\phi^i - u^i\phi_x^i \right)\big|_0^{\ell_i} d\tau.$$

If we add these equalities we get, in view of the boundary conditions (2.12)-(2.15),

$$\int_0^t h\partial_n\phi^1(\tau, \mathbf{v}_1)d\tau = \sum_{i=1}^M \int_0^{\ell_i} \left(u^i(t, x)\phi_t^i(t, x) - u_t^i(t, x)\phi^i(t, x) \right) dx,$$

and this equality means that

$$\langle \partial_n\phi^1(t, \mathbf{v}_1), h \rangle_{L^2(0,t)} = \langle \bar{u}(t), \bar{\phi}_t(t) \rangle_{H \times H} - \langle \bar{u}_t(t), \bar{\phi}(t) \rangle_{V' \times V}. \tag{2.31}$$

Consequently we have

$$\langle \mathbf{A}\bar{\phi}, h \rangle_{L^2(0,t)} = \langle \mathbf{B}h, \bar{\phi} \rangle_{(H \times V') \times (H \times V)}.$$

That is, $\mathbf{B}h = \mathbf{A}_t^* h$ for $h \in C^1([0, t])$. Taking into account that the operator $\mathbf{A}_t^*$ is continuous and that $C^1([0, t])$ is dense in $L^2(0, t)$, we can ensure that $\mathbf{A}_t^*$ coincides with the extension of $\mathbf{B}$ to $L^2(0, t)$.

This fact gives sense to the equality (2.30). In the classical case $h \in C^1([0, t])$, $(\bar{u}_0, \bar{u}_1) \in (H \times V')$, $u_0^i, u_1^i \in C^1([0, \ell_i])$, formula (2.30) simply expresses the fact that the solution of the non-homogeneous problem with initial state $(\bar{u}_0, \bar{u}_1)$ can be represented as the sum of the solution of the homogeneous problem with initial state $(\bar{u}_0, \bar{u}_1)$ and the solution of the non-homogeneous problem with null initial state $(\bar{0}, \bar{0})$. This fact is an immediate consequence of the linearity of the system (2.11)-(2.16).

Finally, for every $h \in L^2(0, T)$ the solution $\bar{u}$ of (2.11)-(2.16) defined by (2.30) has the property

$$\bar{u} \in C([0, T] : H) \bigcap C^1([0, T] : V'). \tag{2.32}$$

Indeed, in view of (2.31) and the estimate (2.29) it follows that

$$\bar{u} \in L^{\infty}(0, T : H) \bigcap W^{1,\infty}(0, T : V'), \tag{2.33}$$

together with the estimate

$$||\bar{u}||_{L^{\infty}(0,T:H)} + ||\bar{u}_t||_{L^{\infty}(0,T:V')} \leq C[||(\bar{u}_0, \bar{u}_1)||_{H \times V'} + ||h||_{L^2(0,T)}]. \tag{2.34}$$

The time continuity of the solution in (2.32) is a consequence of a density argument, (2.33), (2.34) and the fact that, for smooth data, the solution $\bar{u}$ is smooth as well.

2.3 The Control Problem

The control problem in time T consists in determining for which initial states it is possible to choose the controls $h_j \in L^2(0,T)$, $j = 1, ..., r$, such that the system reaches the equilibrium position at time T. Depending on how strict we are on requiring the state to reach equilibrium, several notions or degrees of controllability may be distinguished.

2.3.1 Basic Definitions

More precisely,

Definition 2.1. Let $T > 0$. We say that the initial state $(\bar{u}_0, \bar{u}_1) \in H \times V'$, is **controllable in time** T, if there exist functions $h_j \in L^2(0,T)$, $j = 1, ..., r$, such that the solution of (2.11)-(2.16) with initial state $(\bar{u}_0, \bar{u}_1)$ satisfies

$$\bar{u}|_{t=T} = \bar{u}_t|_{t=T} = \bar{0}.$$

When for every $\varepsilon > 0$ there exist controls h_i^ε such that the corresponding solutions $\bar{u}^\varepsilon$ verify

$$||(\bar{u}^\varepsilon|_T, \bar{u}_t^\varepsilon|_T)||_{H \times V'} < \varepsilon,$$

it is said that $(\bar{u}_0, \bar{u}_1)$ is **approximately controllable in time** T.

Remark 2.2. Sometimes, under the conditions of the Definition 2.1 it is also said that $(\bar{u}_0, \bar{u}_1)$ is exactly controllable.

The following definition classifies systems of the form (2.11)-(2.16) according to the answer to the control problem.

Definition 2.3. Let $T > 0$. We say that the set $K \subset H \times V'$ is controllable in time T, if all the initial states $(\bar{u}_0, \bar{u}_1) \in K$ are controllable in time T. Then, we shall say that the system (2.11)-(2.16) is

1) **approximately** controllable in time T if there exists a dense (in $H \times V'$) set K, which is approximately controllable [4] in time T

---

[4] In other words, system (2.11)-(2.16) is approximately controllable if all the initial states $(\bar{u}_0, \bar{u}_1) \in H \times V'$ are approximately controllable.

2) **spectrally** *controllable in time T if the subspace $Z \times Z$ is controllable in time T, where Z is the set of all the finite linear combinations of the eigenfunctions of the operator $-\Delta_G$;*

3) **exactly** *controllable in time T if the whole space $H \times V'$ is controllable in time T.*

Let us note that, due to the linear character of system (2.11)-(2.16), if the set K is controllable, so is the subspace span K of all finite linear combinations of elements of K. Thus, it is more natural to talk of controllable subspaces instead of controllable sets.

Remark 2.4. Due to the linearity and time reversibility of system, if it is exactly controllable, then every initial datum in $H \times V'$ can be driven at time T to any given state in $H \times V'$ and not only to the zero one. Similarly, when the system is approximately controllable, any initial state can be driven to a dense set of final states.

2.3.2 An Equivalent Formulation of the Control Problem

Let us observe first, that the control problem admits an equivalent formulation in terms of operators. Let $\mathbf{P}_T : U \to H \times V'$ be the operator defined by

$$\mathbf{P}_T \bar{h} := (\bar{u}(T), \bar{u}_t(T)),$$

where $\bar{u}$ is the solution of system (2.11)-(2.16) with initial state $(\bar{0}, \bar{0})$.

Let us denote by $\mathcal{W}_T$ the rank of $\mathbf{P}_T$; that is, $\mathcal{W}_T$ is the set of those states that can be reached after a time T starting from the rest state $(\bar{0}, \bar{0})$.

Let us note that the initial state $(\bar{u}_0, -\bar{u}_1) \in H \times V'$ is controllable in time T if, and only if, $(\bar{u}_0, \bar{u}_1) \in \mathcal{W}_T$. This fact is due to the time-reversibility of system, i.e. to the invariance of system (2.11)-(2.16) under the change of variable $t \to T - t$: if $\bar{u}$ is a solution of (2.11)-(2.16) then, $\bar{w}(t) = \bar{u}(T - t)$ is also a solution. Thus, given $(\bar{u}_0, \bar{u}_1) \in \mathcal{W}_T$, if $\bar{u}$ is a solution satisfying

$$(\bar{u}(0), -\bar{u}_t(0)) = (\bar{0}, \bar{0}), \qquad (\bar{u}(T), \bar{u}_t(T)) = (\bar{u}_0, \bar{u}_1)$$

with control $\hat{h}$ then, $\bar{w}(t) = \bar{u}(T - t)$ satisfies

$$(\bar{w}(0), \bar{w}_t(0)) = (\bar{u}_0, \bar{u}_1), \qquad (\bar{w}(T), \bar{w}(T)) = (\bar{0}, \bar{0}).$$

Consequently, to drive $(\bar{u}_0, -\bar{u}_1)$ to $(\bar{0}, \bar{0})$ it is sufficient to choose the control $\hat{h}(T - t)$.

As a consequence, if the initial states $(\bar{u}_0, \bar{u}_1)$ and $(\bar{v}_0, -\bar{v}_1)$ are controllable in time T then it is possible to find a control $\hat{h} \in U$ driving $(\bar{u}_0, \bar{u}_1)$ to $(\bar{v}_0, \bar{v}_1)$. Indeed, it suffices to take $\hat{h} = \hat{h}_1 + \hat{h}_2$, where $\hat{h}_1$, $\hat{h}_2$ are the controls that drive $(\bar{u}_0, \bar{u}_1)$ to $(\bar{0}, \bar{0})$ and $(\bar{0}, \bar{0})$ to $(\bar{v}_0, \bar{v}_1)$, respectively.

Thus, the control problem in time T is reduced to study the rank $\mathcal{W}_T$ of the operator $\mathbf{P}_T$. On the other hand, on the basis of general results of Functional

Analysis (see Theorem 3.4), the space $\mathcal{W}_T$ may be described in terms of the adjoint operator to $\mathbf{P}_T$. This is essentially the Hilbert Uniqueness Method (HUM).

Let us observe now that, according to the definition (2.30) of the solution of (2.11)-(2.16), the adjoint of the operator $\mathbf{P}_T$ coincides with $\mathbf{A}_T$, that is, the adjoint of $\mathbf{P}_T$ is the operator that associates to $(\bar{\phi}_1, -\bar{\phi}_0) \in H \times V$ the vector $\partial_n \bar{\phi}|_{\mathcal{C}} \in U$, whose components are the normal derivatives $\partial_n \phi^i(., \mathbf{v}_j)$, $j = 1, ..., r$, of the solution of the homogeneous system (2.17)-(2.21) with initial state $(\bar{\phi}_0, \bar{\phi}_1)$. That is why the control problem is reduced to the study of properties of the solutions of the homogeneous system (2.17)-(2.21).

On the other hand, $(\bar{u}_0, \bar{u}_1) \in \mathcal{W}_T$, that is, $(\bar{u}_0, \bar{u}_1) = \mathbf{P}_T \bar{h}$ for some $\bar{h} \in U$, if, and only if, for all $(\bar{\phi}_1, -\bar{\phi}_0) \in Z \times Z$ the following equality is satisfied

$$\langle (\bar{u}_0, \bar{u}_1), (\bar{\phi}_1, -\bar{\phi}_0) \rangle_{(H \times V') \times (H \times V)} = \langle \mathbf{P}_T \bar{h}, (\bar{\phi}_1, -\bar{\phi}_0) \rangle_{(H \times V') \times (H \times V)}.$$

Then, $(\bar{u}_0, \bar{u}_1) \in \mathcal{W}_T$ if, and only if,

$$\langle (\bar{u}_0, \bar{u}_1), (\bar{\phi}_1, -\bar{\phi}_0) \rangle_{(H \times V') \times (H \times V)} = \langle \bar{h}, \mathbf{P}_T^*(\bar{\phi}_1, -\bar{\phi}_0) \rangle_U = \langle \bar{h}, \partial_n \bar{\phi}|_{\mathcal{C}} \rangle_U.$$

Accordingly, we have the following characterization of controllability:

Proposition 2.5. *The initial state $(\bar{u}_0, \bar{u}_1) \in H \times V'$ is controllable in time T with controls $\bar{h} = (h_1, ..., h_r) \in U$ if, and only if, for every $(\bar{\phi}_0, \bar{\phi}_1) \in Z \times Z$ the following equality holds*

$$-\langle \bar{u}_0, \bar{\phi}_1 \rangle_H + \langle \bar{u}_1, \bar{\phi}_0 \rangle_{V' \times V} = \sum_{j=1}^r \int_0^T h_j(t) \partial_n \phi^i(t, \mathbf{v}_j) dt, \qquad (2.35)$$

where $\bar{\phi}$ is the solution of system (2.17)-(2.21) with initial state $(\bar{\phi}_0, \bar{\phi}_1)$.

Remark 2.6. The relation (2.35) suggests a minimization algorithm for the construction of the control $\bar{h}$. If we look for the control in the form $\bar{h} = -\partial_n \bar{\psi}|_{\mathcal{C}}$, where $\bar{\psi}$ is a solution of the homogeneous system (2.17)-(2.21), then the equality (2.35) is the Euler equation $I'(\bar{\psi}_0, \bar{\psi}_1) = 0$ corresponding to the quadratic functional $I : V \times H \to \mathbb{R}$ defined by

$$I(\bar{\phi}_0, \bar{\phi}_1) = \frac{1}{2} \int_0^T \sum_{j=1}^r |\partial_n \phi^i(t, \mathbf{v}_j)|^2 dt + \langle \bar{u}_0, \bar{\phi}_1 \rangle - \langle \bar{u}_1, \bar{\phi}_0 \rangle.$$

Therefore, if $(\bar{\psi}_0, \bar{\psi}_1)$ is a minimizer of I, the relation (2.35) will be verified. The functional I is continuous and convex. So, in order to guarantee the controllability of an initial state $(\bar{u}_0, \bar{u}_1) \in H \times V'$ it is sufficient that I be coercive. This is the central idea of the Hilbert Uniqueness Method (HUM) introduced by J.-L. Lions in [77]. In Chapter 3 we will describe in detail this technique.

2.4 A Controllability Theorem and its Limitations

A natural starting point for the study of the control problem for a network of strings is the following theorem due to G. Schmidt [108].

Theorem 2.7 (Schmidt, [108]).
If G is a tree (it does not contain closed paths) and the set $\mathcal{C}$ contains all the exterior nodes, except at most one, then system (2.11)-(2.16) is exactly controllable in any time $T \geq T^*$, where T^* is twice the length of the largest simple path connecting the uncontrolled node with the controlled ones.

The proof of this theorem is rather simple. The main ingredient is the possibility of representing the solutions of the 1-d wave equation at every string by means of the D'Alembert formula. In Section 3.4.1 of Chapter 3 we describe the proof for the case of a network formed by three strings with two controlled nodes. There we also explain how to proceed in the case of arbitrary trees. Both facts, the tree structure and that all the exterior nodes, except at most one, are controlled, play an essential role in the proof.

The conditions of Theorem 2.7 seem to be very strong: a high number of controls and a simple topological configuration of the graph are required. The question on whether these conditions may be weakened arises naturally. Can the system (2.11)-(2.16) be exactly controllable when there are more than two uncontrolled exterior nodes or when G contains circuits, at least for some values of the lengths of the strings? It turns out that in both cases the answer is negative. In Chapter 6 (Section 6.3) we will prove the following result:

Theorem 2.8. *If G is a tree and there are at least two uncontrolled nodes, then system (2.11)-(2.16) is not exactly controllable whatever $T > 0$ is, i.e. there exist initial states which are not controllable in any finite time T.*

This fact adds particular interest to Theorem 2.7, which turns out to be sharp in what concerns exact controllability.

In these notes we will mainly study networks of strings, which are controlled from their exterior nodes and which do not verify the conditions of Theorem 2.7. Consequently, we only expect the controllability of the system to hold in strict subspaces of $H \times V'$.

3

Some Useful Tools

3.1 D'Alembert Formula and Boundary Observability of the $1 - d$ Wave Equation

In this section we write the D'Alembert formula for the solutions of the $1 - d$ wave equation in a convenient way, to later study of the propagation of solutions along networks. This allows, in particular, to prove observability properties of the solutions of the $1 - d$ wave equation when measurements are done only on one end-point of the string.

3.1.1 D'Alembert Formula

Let us assume that the function $u(t, x)$ satisfies the $1 - d$ wave equation in $\mathbb{R} \times \mathbb{R}$. Then, for every $t_* \in \mathbb{R}$ the function u may be expressed by means of the D'Alembert formula

$$u(t, x) = \frac{1}{2} \left(u(t_*, x + t - t_*) + u(t_*, x - t + t_*) \right) + \frac{1}{2} \int_{x - t + t_*}^{x + t - t_*} u_t(t_*, \xi) d\xi.$$

$$\text{(3.1)}$$

In account of the symmetry of the wave equation with respect to the variables x, t, the formula (3.1) is also valid if we change their role. Thus, if $u(t, x)$ satisfies the $1 - d$ wave equation in $\mathbb{R} \times [0, \ell]$ then, for every $a \in [0, \ell]$, $u(t, x)$ may be expressed by the sidewise formula

$$u(t, x) = \frac{1}{2} \left(u(t + x - a, a) + u(t - x + a, a) \right) + \frac{1}{2} \int_{t - x + a}^{t + x - a} u_x(\tau, a) d\tau. \quad \text{(3.2)}$$

From (3.2), after derivation, we obtain the equalities

$$u_x(t,x) = \frac{1}{2}\left(u_t(t+x-a,a) - u_t(t-x+a,a)\right) + \tag{3.3}$$

$$+ \frac{1}{2}\left(u_x(t+x-a,a) + u_x(t-x+a,a)\right),$$

$$u_t(t,x) = \frac{1}{2}\left(u_t(t+x-a,a) + u_t(t-x+a,a)\right) + \tag{3.4}$$

$$+ \frac{1}{2}\left(u_x(t+x-a,a) - u_x(t-x+a,a)\right).$$

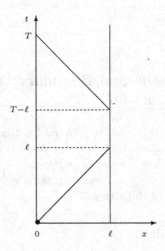

Fig. 3.1. Region of application of the D'Alembert formula

If we denote

$$G(t) := u_t(t,0), \quad F(t) := u_x(t,0), \qquad \widehat{G}(t) := u_t(t,\ell), \quad \widehat{F} := u_x(t,\ell),$$

then formulas (3.3)-(3.4) for $x = \ell$, $a = 0$ may be written as

$$\widehat{F} = \ell^+ F + \ell^- G, \qquad \widehat{G} = \ell^- F + \ell^+ G, \tag{3.5}$$

and, for $x = 0$, $a = \ell$,

$$F = \ell^+ \widehat{F} - \ell^- \widehat{G}, \qquad G = -\ell^- \widehat{F} + \ell^+ \widehat{G}, \tag{3.6}$$

where ℓ^+, ℓ^- are the linear operators acting over time-dependent functions f according to

$$\ell^\pm f(t) := \frac{f(t+\ell) \pm f(t-\ell)}{2}. \tag{3.7}$$

Let us remark that the formulas (3.5) and (3.6) express the relation between the traces of u_t and u_x in the extremes of the interval $[0,\ell]$. Obviously, (3.6) is the inverse relation to (3.5).

3.1.2 Boundary Observability of the $1 - d$ Wave Equation

The following proposition contains a very useful result on the observability of $1 - d$ waves from the boundary. It will be frequently used in what follows.

Proposition 3.1. *If $u(t, x)$ satisfies the wave equation*

$$u_{tt} = u_{xx} \quad in \ \mathbb{R} \times [0, \ell]$$

then

$$\mathbf{E}_u(t) \leq \frac{1}{4} \int_{t-\ell}^{t+\ell} \left(|u_x(\tau, 0)|^2 + |u_t(\tau, 0)|^2 \right) d\tau.$$

Remark 3.2. Note that no boundary conditions are required for this inequality to hold.

Proof. In view of (3.3)-(3.4), it holds

$$\mathbf{E}_u(t) = \frac{1}{8} \int_0^\ell \left\{ |u_t(t+x, 0) - u_t(t-x, 0) + u_x(t+x, 0) + u_x(t-x, 0)|^2 + \right.$$
$$\left. + |u_t(t+x, 0) + u_t(t-x, 0) + u_x(t+x, 0) - u_x(t-x, 0)|^2 \right\} dx$$
$$\leq \frac{1}{4} \int_0^\ell \left\{ |u_t(t+x, 0)|^2 + |u_t(t-x, 0)|^2 + |u_x(t+x, 0)|^2 + |u_x(t-x, 0)|^2 \right\} dx$$
$$= \frac{1}{4} \int_{t-\ell}^t \left\{ |u_t(\tau, 0)|^2 + |u_x(\tau, 0)|^2 \right\} d\tau + \frac{1}{4} \int_t^{t+\ell} \left\{ |u_t(\tau, 0)|^2 + |u_x(\tau, 0)|^2 \right\} d\tau$$
$$= \frac{1}{4} \int_{t-\ell}^{t+\ell} \left(|u_x(\tau, 0)|^2 + |u_t(\tau, 0)|^2 \right) d\tau.$$

Proposition 3.3. *For all $\ell > 0$, a, $b \in \mathbb{R}$ the operators ℓ^+, ℓ^- are continuous from $L^2[a - \ell, b + \ell]$ into $L^2[a, b]$.*

Proof. In addition, we will prove that the norm of the operators $\ell^\pm$, considered as elements of $\mathcal{L}(L^2[a - \ell, b + \ell], L^2[a, b])$, is not greater than one. In fact,

$$\int_a^b |\ell^\pm f(t)|^2 dt = \frac{1}{4} \int_a^b |f(t + \ell) \pm f(t - \ell)|^2 dt$$
$$\leq \frac{1}{2} \int_a^b |f(t + \ell)|^2 dt + \frac{1}{2} \int_a^b |f(t - \ell)|^2 dt$$
$$\leq \frac{1}{2} \int_{a+\ell}^{b+\ell} |f(t)|^2 dt + \frac{1}{2} \int_{a-\ell}^{b-\ell} |f(t)|^2 dt \leq \int_{a-\ell}^{b+\ell} |f(t)|^2 dt.$$

3.2 The Hilbert Uniqueness Method (HUM): Reduction to an Observability Problem.

3.2.1 Description of the Method

In this section we describe the main tool used along these notes for the study of control problems: The *Hilbert Uniqueness Method* (HUM)[1], which allows to reduce the control problem to the study of observability properties of the solutions of the homogeneous system, without controls.

We illustrate the application of HUM for system (2.11)-(2.16). We do it in an abstract setting that allows avoiding the difficulties related to the notations. This allows using it to address other control problems: when equation (2.11) is replaced by the Schrödinger or heat equations, or when the boundary conditions or the choice of the controls are different.

The starting point of HUM consists in reducing the control problem to the identification of the image of a continuous linear operator as it has been described in Section 2.3. This is done with the aid on the following general result of Functional Analysis: if E and F are Hilbert spaces and $\mathbf{A} : F \to E$ is a continuous linear operator with adjoint $\mathbf{A}^* : E' \to F$ (we have identified F and F' through the Riesz-Fréchet isometry) then

Theorem 3.4. *If $\mathbf{A}^*$ is injective then the image of $\mathbf{A}$ coincides with the set*

$$M = \{u \in E : \quad \exists C_u > 0 \text{ such that } |\langle \phi, u \rangle_{E' \times E}| \leq C_u \|\mathbf{A}^*\phi\|_F \ \forall \ \phi \in E'\}.$$

Proof. We will show first that Im $\mathbf{A} \subset M$. If $u \in$ Im $\mathbf{A}$, that is, $u = \mathbf{A}p$ for $p \in F$ then, for all $\phi \in E'$

$$|\langle \phi, u \rangle_{E' \times E}| = |\langle \phi, \mathbf{A}p \rangle_{E' \times E}| = |\langle \mathbf{A}^*\phi, p \rangle_F| \leq \|\mathbf{A}^*\phi\|_F \|p\|_F,$$

and thus $u \in M$ with $C_u = \|p\|_F$.

The inclusion $M \subset$ Im $\mathbf{A}$ is more delicate. Since $\mathbf{A}^*$ is injective, $\mathbf{A}^*\phi = 0$ if, and only if, $\phi = 0$. Consequently, the function $\|\phi\|_{\mathbf{A}} = \|\mathbf{A}^*\phi\|_F$ is a norm in E'. Let $H_{\mathbf{A}}$ be the completion of E' with respect to that norm. This means that there exists an isometry $\kappa : (E', \|.\|_{\mathbf{A}}) \to H_{\mathbf{A}}$ such that $\kappa(E')$ is dense in $H_{\mathbf{A}}$. If we identify E' and $\kappa(E')$ through κ, it holds $E' \subset H_{\mathbf{A}}$. This imbedding is dense and continuous. Indeed, since $\mathbf{A}^*$ is bounded,

$$\|\phi\|_{\mathbf{A}} = \|\mathbf{A}^*\phi\|_F \leq C\|\phi\|_{E'}.$$

For $u \in M$ and $\phi \in E'$ we will denote by $\langle \phi, u \rangle$ the imagine by ϕ of the linear and continuous functional obtained by extending ϕ to M by continuity: if the sequence $(\phi_n) \subset E'$ converges to ϕ in $H_{\mathbf{A}}$ then

$$|\langle \phi_n, u \rangle_{E' \times E} - \langle \phi_m, u \rangle_{E' \times E}| = |\langle \phi_n - \phi_m, u \rangle_{E' \times E}| \leq C_u \|\mathbf{A}^*(\phi_n - \phi_m)\|_F$$
$$= C_u \|\phi_n - \phi_m\|_{\mathbf{A}},$$

---

[1] The name of this method is due to its author J.-L. Lions (see [77], [79], [78]).

and therefore $\langle \phi_n, u \rangle$ is a Cauchy sequence in $\mathbb{R}$ ($(\langle \phi_n \rangle)$ is convergent), and thus it also converges. Now define $\langle \phi, u \rangle = \lim_{n \to \infty} \langle \phi_n, u \rangle$. Passing to the limit in the relations $|\langle \phi_n, u \rangle| \leq C_u \|\phi_n\|_{\mathbf{A}}$ it holds

$$|\langle \phi, u \rangle| \leq C_u \|\phi\|_{\mathbf{A}}. \tag{3.8}$$

We deduce that the mapping $\langle ., u \rangle : H_{\mathbf{A}} \to \mathbb{R}$ is linear and continuous.

Thus, let us consider now the functional $I : H_{\mathbf{A}} \to \mathbb{R}$ defined by

$$I(\phi) = \frac{1}{2} \|\phi\|_{\mathbf{A}}^2 - \langle \phi, u \rangle,$$

which is clearly continuous and convex. Once again, in view of (3.8), I is also coercive:

$$|I(\phi)| \geq \frac{1}{2} \|\phi\|_{\mathbf{A}}^2 - |\langle \phi, u \rangle| \geq \frac{1}{2} \|\phi\|_{\mathbf{A}}^2 - C_u \|\phi\|_{\mathbf{A}} \to \infty$$

as $\|\phi\|_{\mathbf{A}} \to \infty$. Then there exists a minimizer $\hat{\phi} \in H_{\mathbf{A}}$ that, taking into account that I is differentiable, satisfies the Euler equation $I'\hat{\phi} = 0$ and that is

$$\langle \phi, \hat{\phi} \rangle_{\mathbf{A}} = \langle \mathbf{A}^*\phi, \mathbf{A}^*\hat{\phi} \rangle_F = \langle \phi, u \rangle \qquad \text{for all } \phi \in H_{\mathbf{A}}.$$

In particular, for $\phi \in E'$,

$$\langle \phi, \mathbf{A}\mathbf{A}^*\hat{\phi} \rangle_{E' \times E} = \langle \mathbf{A}^*\phi, \mathbf{A}^*\hat{\phi} \rangle_F = \langle \phi, u \rangle_{E' \times E}.$$

This means that

$$u = \mathbf{A}\mathbf{A}^*\hat{\phi} \in \operatorname{Im} \mathbf{A}. \tag{3.9}$$

Remark 3.5. Proceeding in a similar way as in the proof of the previous theorem it may be shown that it is possible to identify "by continuity" $H_{\mathbf{A}}'$ with a subspace of E. In such case,

$$H_{\mathbf{A}}' = \{u \in E : \exists C_u > 0 \text{ s. t. } | < \phi, u >_{E' \times E} | \leq C_u \|\mathbf{A}^*\phi\|_F \quad \forall \phi \in E'\}.$$

Then, from the previous theorem it follows that $\operatorname{Im} \mathbf{A} = H_{\mathbf{A}}'$.

Remark 3.6. In general, the equality $\overline{\operatorname{Im} \mathbf{A}} = (\ker \mathbf{A}^*)^\perp$ holds. From this fact, it holds that $\operatorname{Im} \mathbf{A}$ is dense in E if, and only if, $\mathbf{A}^*$ is injective. Consequently, Theorem 3.4 provides a description of $\operatorname{Im} \mathbf{A}$ whenever it is dense in E. On the other hand, the injectivity of $\mathbf{A}^*$ is equivalent to the fact that the equation $\mathbf{A}^*\phi = v$ has at most one solution. This is the uniqueness property to which the term "uniqueness" in HUM refers.

HUM provides the control of minimal norm, and this by minimizing a suitable quadratic, convex and coercive functional in a Hilbert space.

Let us assume now that W is a Hilbert space such that $W \subset E$ with continuous and dense embedding. This allows to extend by continuity the linear and continuous functionals defined in W to E in a unique way. Consequently, we can view E' as a subspace of W'.

The following result is very useful in order to characterize subspaces of Im **A**.

Corollary 3.7. *The subspace W is contained in the image of the operator **A** if, and only if, there exists a constant $C > 0$ such that*

$$\|\phi\|_{W'} \leq C\|\mathbf{A}^*\phi\|_F, \tag{3.10}$$

for every $\phi \in E'$. In such case, for every $u \in W$ there exists $p \in F$ such that

$$\|p\|_F \leq 2C\|u\|_W. \tag{3.11}$$

Proof. Consider the set

$$\Gamma = \{\phi \in E' : \|\mathbf{A}^*\phi\|_F = 1\} \subset E'.$$

Let us observe that the existence of a constant $C > 0$ such that

$$\|\phi\|_{W'} \leq C\|\mathbf{A}^*\phi\|_F, \tag{3.12}$$

for all $\phi \in E'$ means that Γ is bounded in W'.

On the other hand, the embedding $W \subset$ Im **A** is equivalent to the fact that Γ is weakly bounded in W'. Indeed, according to Theorem 3.4, $W \subset$ Im **A** if an only if, for every $u \in W$ there exists a constant C_u such that

$$| < \phi, u >_{W' \times W} | = | < \phi, u >_{E' \times E} | \leq C_u \|\mathbf{A}^*\phi\|_F, \tag{3.13}$$

for every $\phi \in E'$. Consequently, if $W \subset$ Im **A** then, for all $\phi \in \Gamma$, $u \in W'$,

$$| < \phi, u >_{W' \times W} | \leq C_u, \tag{3.14}$$

that is, Γ is weakly bounded. Conversely, if the inequality (3.14) is verified and $\psi \in E'$ then, choosing $\phi = \psi/\|\mathbf{A}^*\psi\|_F \in \Gamma$ ($\|\mathbf{A}^*\psi\|_F \neq 0$ as $\mathbf{A}^*$ is injective) it holds

$$| < \psi, u >_{W' \times W} | = \|\mathbf{A}^*\psi\|_F | < \phi, u >_{W' \times W} | \leq C_u \|\mathbf{A}^*\psi\|_F, \tag{3.15}$$

and then $u \in$ Im $\mathbf{A}^*$.

Finally, it suffices to recall the fact that the properties of being bounded and weakly bounded coincide in Hilbert spaces [2].

In order to prove (3.11) it suffices to choose for $u \in W$, the element $p \in F$ obtained in the proof of Theorem 3.4, that is, $p = \mathbf{A}^*\hat{\phi}$, where $\hat{\phi}$ is a minimizer of the functional

---

[2] This is an immediate consequence of the Banach-Steinhaus theorem and the reflexivity of Hilbert spaces

$$I(\phi) = \frac{1}{2}\|\phi\|_{\mathbf{A}}^2 - \langle \phi, u \rangle.$$

Then we have,

$$0 = I(0) \geq I(\hat{\phi}) = \frac{1}{2}\|\hat{\phi}\|_{\mathbf{A}}^2 - \langle \hat{\phi}, u \rangle$$

and,

$$\|\hat{\phi}\|_{\mathbf{A}}^2 \leq 2\langle \hat{\phi}, u \rangle \leq 2\|\phi\|_{W'}\|u\|_W \leq 2C\|\hat{\phi}\|_{\mathbf{A}}\|u\|_W.$$

Finally, since $\|p\|_F = \|\hat{\phi}\|_{\mathbf{A}}$, it holds

$$\|p\|_F \leq 2C\|u\|_W.$$

Remark 3.8. In particular, Im $\mathbf{A} = E$, that is, the operator $\mathbf{A}$ is surjective if, and only if there exists a constant $C > 0$ such that

$$\|\phi\|_{E'} \leq C\|\mathbf{A}^*\phi\|_F, \tag{3.16}$$

for every $\phi \in E'$. This condition is equivalent to the continuity of $(\mathbf{A}^*)^{-1}$.

Remark 3.9. Due to the continuity of $\mathbf{A}^*$, it is sufficient to prove the inequality (3.10) for a dense subspace of E'.

Remark 3.10. Inequalities of the form (3.16) are called generically **observability inequalities**.

Let us see now another possible way of constructing subspaces of Im $\mathbf{A}$, which will be frequently used in what follows. Assume that $\mathbf{B} : E' \rightarrow E'$ is a continuous operator, whose image is dense in E' and verifies the properties:

1) There exists a constant $C > 0$ such that, for every $\phi \in E'$,

$$\|\mathbf{B}\phi\|_{E'} \leq C\|\mathbf{A}^*\phi\|_F,$$

for all $\phi \in E'$.

2) If $\mathbf{B}$ is not injective then, neither is $\mathbf{A}^*$; that is, if there exists $\phi \in E' \setminus \{0\}$ such that $\mathbf{B}\phi = 0$ then there exists $\psi \in E' \setminus \{0\}$ such that $\mathbf{A}^*\psi = 0$.

Let us note that an operator $\mathbf{B}$ with the properties indicated above is injective if, and only if, $\mathbf{A}^*$ is injective. Consequently, it holds that the subspace Im $\mathbf{A}$ is dense in E if, and only if, $\mathbf{B}$ is injective.

In that case, property 1 corresponds to the continuity of $\mathbf{B} \circ (\mathbf{A}^*)^{-1}$. Moreover, if in addition $\mathbf{B}$ were surjective, then, according to the Banach open mapping theorem, its inverse $\mathbf{B}^{-1}$ would also be continuous and the same for $(\mathbf{A}^*)^{-1}$; so we would have Im $\mathbf{A} = E$.

This cannot be asserted if $\mathbf{B}$ is not surjective. However, it is true for some smaller subspace:

Proposition 3.11. *If* **B** *is a continuous operator with dense image having the property 1, then* Im $\mathbf{B}^* \subset$ Im $\mathbf{A}$, *where* $\mathbf{B}^*$ *is the adjoint operator to* $\mathbf{A}$.

Proof. If $u \in$ Im $\mathbf{B}^*$, that is, $u = \mathbf{B}^* v$ then

$$\langle u, \phi \rangle_{E \times E'} = \langle \mathbf{B}^* v, \phi \rangle_{E \times E'} = \langle v, \mathbf{B}\phi \rangle_{E \times E'} \leq \|v\|_E \|\mathbf{B}\phi\|_{E'} \leq C \|v\|_E \|\mathbf{A}^*\phi\|_F,$$

and so the assertion follows from Theorem 3.4.

Property 2 guarantees that the previous result is exact in the sense that it provides a dense subspace in Im $\mathbf{A}$ whenever such a subspace exists. Unfortunately that subspace does not necessarily coincide with the image of $\mathbf{A}$, it may be smaller.

In this book we will use the results described above in the following particular setting. Let H be a separable Hilbert space and $\{\theta_n\}_{n \in \mathbb{N}}$ an orthonormal basis of H. Let us denote by Φ the set of all the formal linear combinations $\bar{X} = \sum_{n \in \mathbb{N}} x_n \theta_n$, $x_n \in \mathbb{R}$, and Z the set of finite linear combinations.

Let (α_n), (β_n) be sequences of real numbers different from zero and define the Hilbert space

$$E := \left\{ (\bar{X}, \bar{Y}) \in \Phi \times \Phi : \quad \|(\bar{X}, \bar{Y})\|_E^2 := \sum_{n \in \mathbb{N}} (\alpha_n^2 x_n^2 + \beta_n^2 y_n^2) < \infty \right\}$$

endowed with the norm $\|.\|_E$. Then, the dual of E may be identified with the space

$$E' = \left\{ (\bar{X}, \bar{Y}) \in \Phi \times \Phi : \quad \|(\bar{X}, \bar{Y})\|_{E'}^2 := \sum_{n \in \mathbb{N}} (\alpha_n^{-2} x_n^2 + \beta_n^{-2} y_n^2) < \infty \right\}$$

endowed with the norm $\|.\|_{E'}$.

Let us consider as before the linear and continuous operator $\mathbf{A} : F \to E$ with injective adjoint $\mathbf{A}^*$. Let now (c_n) be another sequence verifying $c_n \geq c$ for some $c > 0$ and define the space

$$W := \left\{ (\bar{X}, \bar{Y}) \in \Phi \times \Phi : \quad \|(\bar{X}, \bar{Y})\|_W^2 := \sum_{n \in \mathbb{N}} c_n^2 (\alpha_n^2 x_n^2 + \beta_n^2 y_n^2) < \infty \right\} \subset E.$$

Then the results of Corollary 3.7 allow us to assert that

Proposition 3.12. $W \subset$ Im $\mathbf{A}$ *if, and only if, there exists a constant* $C > 0$ *such that*

$$\|(\bar{X}, \bar{Y})\|_{W'}^2 := \sum_{n \in \mathbb{N}} c_n^{-2} (\alpha_n^{-2} x_n^2 + \beta_n^{-2} y_n^2) \leq C \|\mathbf{A}^*(\bar{X}, \bar{Y})\|_F^2, \qquad (3.17)$$

for all $\bar{X}, \bar{Y} \in Z$, *that is, for all finite linear combinations* (x_n), (y_n).

Clearly, if the inequality (3.17) holds, then Im $\mathbf{A}$ contains the subspace $Z \times Z$ of all the finite linear combinations. This is the spectral controllability property.

Let us now analyze what is the minimal condition to guarantee spectral controllability. Observe that, due to the linearity of $\mathbf{A}$, $Z \times Z \subset \operatorname{Im} \mathbf{A}$ if, and only if, $(\bar{\theta}_n, \bar{0})$ and $(\bar{0}, \bar{\theta}_n)$ belong to Im $\mathbf{A}$ for every $n \in \mathbb{N}$. According to Theorem 3.4, the latter fact is equivalent to the existence, for every $n \in \mathbb{N}$, of constants C_n^1, $C_n^2 > 0$ such that

$$| < (\bar{X}, \bar{Y}), (\bar{\theta}_n, \bar{0}) >_{E' \times E} | \leq C_n^1 \|\mathbf{A}^*(\bar{X}, \bar{Y})\|_F,$$

$$| < (\bar{X}, \bar{Y}), (\bar{0}, \bar{\theta}_n) >_{E' \times E} | \leq C_n^2 \|\mathbf{A}^*(\bar{X}, \bar{Y})\|_F.$$

It suffices now to note that

$$< (\bar{X}, \bar{Y}), (\bar{\theta}_n, \bar{0}) >_{E' \times E} = x_n, \qquad < (\bar{X}, \bar{Y}), (\bar{0}, \bar{\theta}_n) >_{E' \times E} = y_n$$

to conclude:

Proposition 3.13. *$Z \times Z \subset \operatorname{Im} \mathbf{A}$ if, and only if, for every $n \in \mathbb{N}$ there exists a constant $C_n > 0$ such that*

$$|x_n| + |y_n| \leq C_n \|\mathbf{A}^*(\bar{X}, \bar{Y})\|_F,$$

for all $\bar{X}$, $\bar{Y} \in Z$.

Remark 3.14. Note that (3.13) implies (3.17) for a suitable sequence (c_n). However, (3.13) in itself does not give any information on how the weights (c_n) may degenerate as n tends to infinty and, consequently, does not suffice to identify the norm $\| \cdot \|_{W'}$ in (3.17).

3.2.2 Application to the Control of Networks

Let us apply now the previous results to the control of the network. From Theorem 3.4 the following holds immediately:

Corollary 3.15. *The initial state $(\bar{u}_0, \bar{u}_1) \in H \times V'$ is controllable in time T if, and only if, there exists a constant $C > 0$ such that*

$$C \int_0^T \sum_{j=1}^r |\partial_n \phi^i(t, \mathbf{v}_j)|^2 dt \geq \left| \langle \bar{u}_0, \bar{\phi}_1 \rangle_H - \langle \bar{u}_1, \bar{\phi}_0 \rangle_{V' \times V} \right|^2$$

for every solution $\bar{\phi}$ of system (2.17)-(2.21) with initial state $(\bar{\phi}_0, \bar{\phi}_1) \in Z \times Z$.

It is interesting to point out how formula (3.9), obtained in the proof of Theorem 3.4, provides an algorithm for the construction of the control $\bar{h}$ that drives the controllable state $(\bar{u}_0, \bar{u}_1) \in H \times V'$ to $(\bar{0}, \bar{0})$ in time T: we should solve the extremal problem

$$I(\Psi^*) = \min_W I(\Psi) \tag{3.18}$$

for the functional

$$I(\Psi) = \frac{1}{2} \int_0^T \sum_{j=1}^r |\partial_n \phi^i(t, \mathbf{v}_j)|^2 dt + \langle \bar{u}_0, \bar{\phi}_1 \rangle - \langle \bar{u}_1, \bar{\phi}_0 \rangle$$

over the space W, which is the completion of $Z \times Z$ with the norm

$$\|(\bar{\phi}_0, \bar{\phi}_1)\|_W = \left[\int_0^T \sum_{j=1}^r |\partial_n \phi^i(t, \mathbf{v}_j)|^2 dt \right]^{\frac{1}{2}}$$

and ϕ is the solution of (2.17)-(2.21) with initial state $\Psi = (\bar{\phi}_0, \bar{\phi}_1)$.

Let $\Psi^* = (\bar{\varphi}_0^*, \bar{\varphi}_1^*)$ be the solution of the minimization problem (3.18). Next, we solve the homogeneous system (2.17)-(2.21) with initial data $(\bar{\varphi}_0^*, \bar{\varphi}_1^*)$. Let $\bar{\phi}$ be the corresponding solution. The control is then the trace $\partial_n \bar{\phi}|_e$ of this solution.

Besides, from Remarks 3.6 and 3.8 it follows

Corollary 3.16. *System (2.11)-(2.16) is approximately controllable in time T if, and only if, the following unique continuation property is verified*

$$\partial_n \phi^i(t, \mathbf{v}_j) = 0, \ j = 1, ..., r, \quad a. \ e. \ t \in [0, T], \ \Rightarrow \ (\bar{\phi}_0, \bar{\phi}_1) = (\bar{0}, \bar{0}).$$

Moreover, all the initial states $(\bar{u}_0, \bar{u}_1) \in H \times V'$ are exactly controllable in time T if, and only if, there exists a constant $C > 0$ such that

$$C \int_0^T \sum_{j=1}^r |\partial_n \phi^i(t, \mathbf{v}_j)|^2 dt \geq \|(\bar{\phi}_0, \bar{\phi}_1)\|_{V \times H}^2 \tag{3.19}$$

for all $(\bar{\phi}_0, \bar{\phi}_1) \in V \times H$.

Inequality (3.19) may be expressed in terms of the Fourier coefficients $(\phi_{0,n}), (\phi_{1,n})$ of the initial data $(\bar{\phi}_0, \bar{\phi}_1)$ as

$$C \int_0^T \sum_{j=1}^r |\partial_n \phi^i(t, \mathbf{v}_j)|^2 dt \geq \sum_{n \in \mathbb{N}} \left(\mu_n \phi_{0,n}^2 + \phi_{1,n}^2 \right). \tag{3.20}$$

Unfortunately, this inequality does not hold for system (2.11)-(2.16), except under the very restrictive conditions on the graph G and the location of the controlled nodes of Theorem 2.7. Therefore, all along this book, we shall deal with situations where inequality (3.19) is not true, that is, such that there exist initial states $(\bar{u}_0, \bar{u}_1) \in V' \times H$, which are not controllable in time T. Consequently, we will only be able to prove weaker inequalities of the type

$$\int_0^T \sum_{j=1}^r |\partial_n \phi^i(t, \mathbf{v}_j)|^2 dt \geq \sum_{n \in \mathbb{N}} c_n^2 \left(\mu_n \phi_{0,n}^2 + \phi_{1,n}^2 \right), \qquad (3.21)$$

with non-vanishing coefficients c_n. This will allow to ensure, according to Proposition 3.12, that the space of initial states $(\bar{u}_0, \bar{u}_1) \in V' \times H$ defined by

$$\sum_{n \in \mathbb{N}} \frac{1}{c_n^2} u_{0,n}^2 < \infty, \qquad \sum_{n \in \mathbb{N}} \frac{1}{c_n^2 \mu_n} u_{1,n}^2 < \infty, \qquad (3.22)$$

is controllable in time T.

From that fact, it would hold, in particular, that the system is spectrally controllable (and then approximately controllable) in time T.

Observe that, if we were able to prove, in addition, that the coefficients c_n in (3.21) verify a uniform inequality of the form

$$c_n^2 \mu_n^\varepsilon \geq C > 0,$$

for some $\varepsilon \in \mathbb{R}$, then, the sequences $(u_{0,n}), (u_{1,n})$ such that

$$\sum_{n \in \mathbb{N}} \mu_n^\varepsilon u_{0,n}^2 < \infty, \qquad \sum_{n \in \mathbb{N}} \mu_n^{\varepsilon-1} u_{1,n}^2 < \infty,$$

would satisfy the inequalities (3.22). This would imply that the space $\mathcal{W}^\varepsilon$ is controllable in time T.

Remark 3.17. Let us assume that $r = 1$ and the inequality (3.21) is verified. If we replace $\bar{\phi}$ by its explicit expression (1.2.23), we obtain

$$\int_0^T |\sum_{k \in \mathbb{N}} \varkappa_k (\phi_{0,k} \cos \lambda_k t + \frac{\phi_{1,k}}{\lambda_k} \sin \lambda_k t)|^2 dt \geq \sum_{k \in \mathbb{N}} c_k^2 (\mu_k \phi_{0,k}^2 + \phi_{1,k}^2), \quad (3.23)$$

where $\varkappa_k = \partial_n \theta_k^1(\mathbf{v}_1)$.

If we define for $k < 0$, $\lambda_k := -\lambda_{|k|}$ and denote $a_k = \left(u_{0,|k|} - i u_{1,|k|}/\lambda_k \right)/2$ for $k \in \mathbb{Z}_*$, the inequality (3.23) becomes

$$\int_0^T |\sum_{k \in \mathbb{Z}_*} \varkappa_{|k|} a_k e^{i\lambda_k t}|^2 dt \geq \sum_{k \in \mathbb{N}} c_k^2 \mu_k |a_k|^2,$$

for every finite sequence (a_k) of complex numbers, satisfying $a_{-k} = \overline{a_k}$.

Let us note, however, that,

$$\frac{1}{2} |\sum_{k \in \mathbb{Z}_*} \varkappa_{|k|} a_k e^{i\lambda_k t}|^2 \leq |\sum_{k>0} \varkappa_{|k|} a_k e^{i\lambda_k t}|^2 + |\sum_{k<0} \varkappa_{|k|} a_k e^{i\lambda_k t}|^2$$

and since

$$\sum_{k<0} \varkappa_{|k|} a_k e^{i\lambda_k t} = \sum_{k>0} \varkappa_k \overline{a_k} e^{-i\lambda_k t} = \overline{\sum_{k>0} \varkappa_k a_k e^{i\lambda_k t}},$$

we obtain that the following inequalities hold

$$\int_0^T |\sum_{k\in\mathbb{N}} \varkappa_k a_k e^{i\lambda_k t}|^2 dt \geq C \sum_{k\in\mathbb{N}} c_k^2 \mu_k |a_k|^2,$$

$$\int_0^T |\sum_{k\in\mathbb{N}} \varkappa_k a_k e^{-i\lambda_k t}|^2 dt \geq C \sum_{k\in\mathbb{N}} c_k^2 \mu_k |a_k|^2,$$

for every finite complex sequence (a_k).

3.3 The Method of Moments

In this section we describe an alternative method for the study of the control problem: the so called *method of moments*. It is useful not only for networks of strings, but also in the study of systems obtained by replacing in (2.11)-(2.16) the wave equation by the heat equation and, in general, by equations, whose solutions may be computed using the method of separation of variables.

3.3.1 Description of the Method

Let H be a Hilbert space and $(\mathbf{a}_n)$ be a sequence of elements of H. Given a sequence $(m_n) \in l^2$, consider the following *problem of moments*: To find an element $v \in H$ such that

$$\langle v, \mathbf{a}_n \rangle_H = m_n, \qquad n \in \mathbb{Z}. \tag{3.24}$$

Problems of moments appear in a natural way in the study of control problems when trying to find the control v that drives an initial state to rest in time T directly from the characterization in Proposition 2.5. In this case, the space H is $L^2(0,T)$ and the sequence $(\mathbf{a}_n)$ is formed by the complex exponentials $\mathbf{a}_n = e^{i\lambda_n t}$. This leads to the problem of moments

$$\int_0^T v(t) e^{i\lambda_n t} dt = m_n, \qquad n \in \mathbb{Z}, \tag{3.25}$$

where the sequence (m_n) depends on the Fourier coefficients of the initial state to be controlled.

Historically, this approach was the first one giving important results on the controllability of partial differential equations. For more details, the reader is referred to the papers [43], [105], [45], [106], [44].

A natural way to search for a solution of (3.24) is to solve first the problem for the sequences of the canonical basis $\bar{\mathbf{e}}^k = \left(\delta_n^k \right)$ of l^2. Here, δ_n^k stands for the Kronecker δ ($\delta_n^k = 1$ if $n = k$ and $= 0$ otherwise). If we denote by v_k the corresponding solutions (assuming that such solutions exists), we have

$$\langle v_k, \mathbf{a}_n \rangle = \delta_n^k \qquad n, k \in \mathbb{Z}.$$

A sequence (v_k) with this property is called *biorthogonal sequence to* the sequence $(\mathbf{a}_n)$ in H. The usefulness of a biorthogonal sequence is immediate: if we choose

$$v = \sum_{k \in \mathbb{N}} m_k v_k, \tag{3.26}$$

we have, at least formally, that, for every n, it holds

$$\langle v, \mathbf{a}_n \rangle = \sum_{k \in \mathbb{N}} m_k \langle v_k, \mathbf{a}_n \rangle = \sum_{k \in \mathbb{N}} m_k \delta_n^k = m_n.$$

Under additional summability conditions on the sequence (m_n), formula (3.26) provides a solution of (3.24):

Proposition 3.18. *If $(v_n) \subset H$ is a biorthogonal sequence to $(\mathbf{a}_n)$ in H then, for every sequence (m_n) such that*

$$\sum_{n \in \mathbb{N}} |m_n| \, \|v_n\|_H < \infty, \tag{3.27}$$

there exists a solution $v \in H$ of (3.24). That solution is given by (3.26).

Proof. It suffices to note that the function v defined by (3.26) belongs to H:

$$\|v\|_H \leq \sum_{n \in \mathbb{N}} |m_n| \, \|v_n\|_H < \infty.$$

Thus, solving a problem of moments with this technique involves two fundamental steps: to determine a biorthogonal sequence and to estimate the norms of its elements. According to Proposition 3.18, if there exists a biorthogonal sequence, we will be able to determine a dense (in l^2) subspace of sequences, defined by (3.27), for which the problem of moments has a solution. In particular, *the existence of a biorthogonal sequence guarantees the solvability of the problem of moments for every finite sequence (m_n).*

As it has been pointed out above, in the study of control problems, the problem of moments (3.25) arises for sequences (λ_n) of complex numbers such that $(\Re \lambda_n)$ is increasing[3]. In this case, a biorthogonal sequence may be constructed in a relatively easy way thanks to the developments of Paley and Wiener [96].

After performing the change of variables $t \to t + A$ with $A = T/2$, problem (3.25) may be written in the symmetric form

---

[3] $\Re z$ denotes the real part of the complex number z.

$$\int_{-A}^{A} \tilde{v}(t)e^{i\lambda_n t}dt = \tilde{m}_n. \tag{3.28}$$

This constitutes a problem of moments in $L^2(-A, A)$.

Let us assume that F is an entire function satisfying:

1) $F \in L^\infty(\mathbb{R})$;
2) F is of exponential type not greater than A: there exist constants $M, A > 0$ such that $|F(z)| \le Me^{A|z|}$ for every $z \in \mathbb{C}$.
3) all the numbers λ_n are simple zeros of F:

$$F(\lambda_n) = 0, \quad F'(\lambda_n) \ne 0.$$

Then, it is easy to see that the functions

$$F_k(z) := \frac{F(z)}{(z - \lambda_k)F'(\lambda_k)} \tag{3.29}$$

satisfy property 2. Besides, it may be shown, using the Phragmén-Lindelöf theorem (see, e.g., Theorem 11, p. 82 in [116]), that there exists a constant $C > 0$ such that for every $k \in \mathbb{N}$,

$$\|F_k\|_{L^2(\mathbb{R})} \le \frac{C}{|F'(\lambda_k)|} \|F\|_{L^\infty(\mathbb{R})}; \tag{3.30}$$

in particular, the functions F_k belong to $L^2(\mathbb{R})$.

Finally, let us observe that $F_k(\lambda_n) = \delta_k^n$.

To apply this technique of the method of moments we need the following fundamental tool, the Paley-Wiener's Theorem:

Theorem 3.19 (Paley and Wiener, [96]). *The function F is the Fourier transform of a function $\varphi \in L^2(\mathbb{R})$ with support contained in the interval $[-A, A]$, that is,*

$$F(z) = \int_{-A}^{A} e^{izt}\varphi(t)dt,$$

if, and only if, F is an entire function of exponential type at most A and $F \in L^2(\mathbb{R})$.

If we apply Theorem 3.19 to the functions F_k defined by (3.29) it holds that there exist functions $v_k \in L^2(-A, A)$ such that

$$F_k(z) = \int_{-A}^{A} e^{izt}v_k(t)dt, \qquad k \in \mathbb{N}.$$

From these inequalities we obtain

$$\int_{-A}^{A} e^{i\lambda_n t} v_k(t) dt = F_k(\lambda_n) = \delta_k^n,$$

and thus, the sequence (v_k) would be biorthogonal to $(e^{i\lambda_n t})$ in $L^2(-A, A)$. By this reason, the function F is called *generating function* for the sequence $(e^{i\lambda_n t})$.

On the other hand, from Plancherel's identity

$$\|v_k\|_{L^2(-A,A)} = \|F_k\|_{L^2(\mathbb{R})}.$$

Consequently, in view of (3.30) there exists a constant $C > 0$ such that for every $k \in \mathbb{N}$

$$\|v_k\|_{L^2(-A,A)} \leq \frac{C}{|F'(\lambda_k)|}. \tag{3.31}$$

Then, if we succeed in constructing a generating function F of the sequence (λ_n), the problem of identifying subspaces of sequences (m_n) for which the problem of moments (3.28) has a solution is reduced to estimate the sequence of norms $|F'(\lambda_k)|$.

Remark 3.20. If it were possible to establish uniform estimates of the form

$$|F'(\lambda_k)| \geq C |\lambda_k|^{-\alpha}, \tag{3.32}$$

then it would hold

$$\|v_k\|_{L^2(-A,A)} \leq C |\lambda_k|^{\alpha}$$

and, according to Proposition 3.18, the problem of moments (3.28) has a solution for every sequence (m_n) satisfying

$$\sum_{n \in \mathbb{N}} |m_n| |\lambda_k|^{\alpha} < \infty.$$

It is useful to characterize subspaces of sequences of the type h^s (as in (2.28)), for which the problem of moments has a solution. Indeed, this provides sufficient conditions on the initial data to be controllable. Let us observe that, if there exists $\gamma \in \mathbb{R}$ such that

$$\sum_{n \in \mathbb{N}} |\lambda_k|^{\gamma} < \infty, \tag{3.33}$$

then, from the Cauchy-Schwarz inequality, it holds

$$\sum_{n \in \mathbb{N}} |m_n| |\lambda_k|^{\alpha} < \sum_{n \in \mathbb{N}} |m_n|^2 |\lambda_k|^{2(\alpha - \frac{\gamma}{2})} \sum_{n \in \mathbb{N}} |\lambda_k|^{\gamma}.$$

Thus, under the conditions (3.32) and (3.33), the problem of moments (3.28) would have a solution for every sequence $(m_n) \in h^{\alpha - \frac{\gamma}{2}}$.

Remark 3.21. It is relatively easy to construct an entire function F vanishing at the elements of the sequence (λ_n) if we have additional information on the numbers λ_n. For instance, if there exists $p \in \mathbb{N}$ such that

$$\sum_{n \in \mathbb{Z}} \frac{1}{|\lambda_n|^p} < \infty,$$

one may take, e.g.,

$$F(z) = \prod_{n \in \mathbb{Z}} \left(\frac{\sin(\pi z / \lambda_n)}{\pi z / \lambda_n} \right)^p,$$

which is a bounded function for $z \in \mathbb{R}$. To guarantee that the zeros of F are all simple is not as easy. However, the true difficulty consists in estimating $F'(\lambda_n)$. In [102] and [76] wide information on this issue may be found. The works [44], [43], [45] constitute good examples of the difficulties involved in the application of this moment problem technique.

The following result due to Russell is very useful when addressing the problem of control for the heat equation. It allows obtaining a biorthogonal sequence to the exponential family appearing in connection with the heat equation from a biorthogonal sequence of the family of exponentials of the wave equation. Essentially, this result is contained in [105], though we state it in a form similar to that in [9, Teorema II.5.20].

Let $(\lambda_n)_{n \in \mathbb{Z}_*}$ be a sequence of real numbers such that $\lambda_{-n} = -\lambda_n$ and $(\varkappa_n)_{n \in \mathbb{Z}_*}$ be a symmetric sequence of complex numbers: $\varkappa_{-n} = \varkappa_n$.

Theorem 3.22 (Russell, [105]).

If there exists a sequence (v_n) biorthogonal to $\left(\varkappa_n e^{i \lambda_n t} \right)_{n \in \mathbb{Z}_*}$ in $L^2(-A, A)$ then, for every $\varepsilon > 0$ there will exist a sequence (w_n) biorthogonal to $\left(\varkappa_n e^{-\lambda_n^2 t} \right)_{n \in \mathbb{N}}$ in $L^2(-\varepsilon, \varepsilon)$. Besides, there exist positive constants C_ε and γ such that

$$\|w_n\|_{L^2(-\varepsilon, \varepsilon)} \leq C_\varepsilon \|v_n\|_{L^2(-A, A)} e^{\gamma |\lambda_n|},$$

for all $n \in \mathbb{N}$.

3.3.2 Application to the Control of Networks

In this section we reformulate the control problem for the network system (2.11)-(2.16) in the context of the method of moments. This will provide an alternative approach for the study of controllability. In what follows we will consider, for simplicity, $r = 1$, that is, the network is controlled from one exterior node only.

According to Proposition 2.5, the initial state $(\bar{u}_0, \bar{u}_1) \in H \times V'$ is controllable in time T if, and only if, there exists $h \in L^2(0, T)$ such that, for every $(\bar{\phi}_0, \bar{\phi}_1) \in Z \times Z$ the following equality holds

$$\int_0^T h(t)\partial_n\phi^1(t, \mathbf{v}_1)dt = \langle \bar{u}_1, \bar{\phi}_0\rangle_{V'\times V} - \langle \bar{u}_0, \bar{\phi}_1\rangle_{H\times H}, \qquad (3.34)$$

where $\bar{\phi}$ is the solution of the homogeneous system (2.17)-(2.21) with initial state $(\bar{\phi}_0, \bar{\phi}_1)$.

Let us observe that, if

$$\bar{\phi}_0 = \sum_{n\in\mathbb{N}} \phi_{0,n}\bar{\theta}_n, \qquad \bar{\phi}_1 = \sum_{n\in\mathbb{N}} \phi_{1,n}\bar{\theta}_n$$

then, from formula (2.23) we have

$$\partial_n\phi^1(t, \mathbf{v}_1) = \sum_{k\in\mathbb{Z}_*} \varkappa_k \left(\phi_{0,k}\cos\lambda_k t + \frac{\phi_{1,k}}{\lambda_k}\sin\lambda_k t \right),$$

where $\varkappa_k = \partial_n\theta_k^1(\mathbf{v}_1)$ is the value of the normal derivative of the eigenfunction θ_k in the controlled node. With this, the condition (3.34) says that the initial state $\bar{u}_0 = \sum_{k\in\mathbb{N}} u_{0,k}\bar{\theta}_k$, $\bar{u}_1 = \sum_{k\in\mathbb{N}} u_{1,k}\bar{\theta}_k$ is controllable in time T with control h if, and only if, for all the finite sequences $(\phi_{0,k}), (\phi_{1,k})$ the following equality is satisfied

$$\int_0^T \sum_{k\in\mathbb{N}} \varkappa_k \left(\phi_{0,k}\cos\lambda_k t + \frac{\phi_{1,k}}{\lambda_k}\sin\lambda_k t \right) h(t)dt = \sum_{k\in\mathbb{N}} (u_{1,k}\phi_{0,k} - u_{0,k}\phi_{1,k}).$$

$$(3.35)$$

By choosing in (3.35) $\phi_{0,k} = 1$, $\phi_{0,j} = 0$ for $j \neq k$ and $\phi_{1,j} = 0$ for every j, what corresponds to the initial data $\bar{\phi}_0 = \bar{\theta}_k$, $\bar{\phi}_1 = \bar{0}$, we obtain

$$\int_0^T \varkappa_k \cos\lambda_k t \; h(t)dt = u_{1,k}. \qquad (3.36)$$

In an analogous way, with $\phi_{1,k} = \lambda_k$, $\phi_{1,j} = 0$ for $j \neq k$ and $\phi_{0,j} = 0$ for all j,

$$\int_0^T \varkappa_k \sin\lambda_k t \; h(t)dt = -\lambda_k u_{0,k}. \qquad (3.37)$$

Naturally, relations (3.36), (3.37) are necessary for (3.35) to be satisfied. Besides, they are sufficient. Indeed, if we multiply (3.36) by $\phi_{0,k}$, (3.37) by $\phi_{1,k}$ and add over a finite set $I \subset \mathbb{N}$ we obtain

$$\int_0^T \sum_{k\in I} \varkappa_k \left(\phi_{0,k}\cos\lambda_k t + \frac{\phi_{1,k}}{\lambda_k}\sin\lambda_k t \right) h(t)dt = \sum_{k\in I} (u_{1,k}\phi_{0,k} - u_{0,k}\phi_{1,k}),$$

and this is (3.35).

Now, combining the equalities (3.36), (3.37) it holds

$$\int_0^T \varkappa_k e^{i\lambda_k t}\, h(t)dt = u_{1,k} - i\lambda_k u_{0,k}, \tag{3.38}$$

$$\int_0^T \varkappa_k e^{-i\lambda_k t}\, h(t)dt = u_{1,k} + i\lambda_k u_{0,k}, \quad k \in \mathbb{N}. \tag{3.39}$$

If we define for $k < 0$, $\lambda_k = -\lambda_{-k}$ then the previous results and (3.38)-(3.39) may be unified in

Proposition 3.23. *The initial state* $(\bar{u}_0, \bar{u}_1)$ *is controllable in time* T *with control* h *if, and only if, the following equalities are verified*

$$\int_0^T \varkappa_{|k|} e^{i\lambda_k t}\, h(t)dt = u_{1,|k|} - i\lambda_k u_{0,|k|} \quad \text{for every } k \in \mathbb{Z}_*, \tag{3.40}$$

Equalities (3.40) constitute a problem of moments for the sequence

$$\left(\varkappa_{|k|} e^{i\lambda_k t} \right)_{k \in \mathbb{Z}_*}.$$

Let us observe that, if h is a real function (what is a natural restriction in the context of control of system (2.11)-(2.16)) any of the relations (3.38)-(3.39) implies (3.36) and (3.37). The reason to split the moment equations into two equalities lies on the fact that the method we will use to solve the problem of moments does not guarantee a priori that the solution h is real. However, if we are able to construct a complex function satisfying (3.40) then, its real part will satisfy (3.36), (3.37). Indeed, it suffices to note that (3.39) may be written as

$$\int_0^T \varkappa_k e^{i\lambda_k t}\, \overline{h(t)}dt = u_{1,k} - i\lambda_k u_{0,k} \quad \text{for } k > 0,$$

from which we obtain, after adding this equality to the first one,

$$\int_0^T \varkappa_k e^{i\lambda_k t}\frac{h(t) + \overline{h(t)}}{2}dt = u_{1,k} - i\lambda_k u_{0,k} \quad \text{for } k > 0.$$

This means, that the real function

$$\hat{h}(t) = \frac{h(t) + \overline{h(t)}}{2}$$

satisfies (3.36) and (3.37).

As a consequence of Proposition 3.23 the following characterization of the spectral controllability property of system (2.11)-(2.16) is obtained:

Proposition 3.24. *System (2.11)-(2.16) is spectrally controllable in time* T *if, and only if, there exists a sequence* $(v_k)_{k \in \mathbb{Z}_*}$ *biorthogonal to* $\left(\varkappa_{|k|} e^{i\lambda_k t} \right)_{k \in \mathbb{Z}_*}$ *in* $L^2(0,T)$.

Proof. The fact that the existence of a sequence biorthogonal to $\left(\varkappa_{|k|}e^{i\lambda_k t}\right)_{k\in\mathbb{Z}_*}$ in $L^2(0,T)$ implies the spectral controllability is immediate: the problem of moments (3.40) would have a solution for any finite sequence $(u_{0,n})$, $(u_{1,n})$ and then, in view of Proposition 3.23, all the initial states in $Z\times Z$ would be controllable in time T.

To see that this condition is also necessary, we assume that system (1.2.11)-(1.2.16) is spectrally controllable and construct a sequence biorthogonal to $\left(\varkappa_{|k|}e^{i\lambda_k t}\right)_{k\in\mathbb{Z}_*}$ in $L^2(0,T)$.

For every $m\in\mathbb{N}$, let $g_m, h_m\in L^2(0,T)$ be the controls that correspond to the initial states $(\bar{\theta}_m,\bar{0})$ and $(\bar{0},\bar{\theta}_m)$, respectively. In such case, according to Proposition 3.23, we have the equalities

$$\int_0^T \varkappa_k e^{i\lambda_k t}\, h_m(t)dt = \delta_{|k|}^m, \qquad \int_0^T \varkappa_k e^{i\lambda_k t}\, g_m(t)dt = -i\lambda_k\delta_{|k|}^m,$$

for $m\in\mathbb{N}$, $k\in\mathbb{Z}_*$.

Let us define the functions

$$v_m = \frac{1}{2}\left(h_{|m|} + \frac{i}{\lambda_{|m|}}g_{|m|}\right), \quad m\in\mathbb{Z}_*. \tag{3.41}$$

We will have

$$\int_0^T \varkappa_k e^{i\lambda_k t}\, v_m(t)dt = \frac{1}{2}\int_0^T \varkappa_k e^{i\lambda_k t}\, h_{|m|}(t)dt + \frac{i}{2\lambda_m}\int_0^T \varkappa_k e^{i\lambda_k t}\, g_{|m|}(t)dt$$

$$= \frac{1}{2}\delta_{|k|}^{|m|} + \frac{\lambda_k}{2\lambda_{|m|}}\delta_{|k|}^{|m|} = \delta_k^m.$$

This means that the sequence $(v_m)_{m\in\mathbb{Z}_*}$ is biorthogonal to $\left(\varkappa_{|k|}e^{i\lambda_k t}\right)_{k\in\mathbb{Z}_*}$.

If we know subspaces of controllable initial states for system (2.11)-(2.16), then it is possible to give more precise information on the biorthogonal sequence constructed in Proposition 3.24:

Proposition 3.25. *If the subspace $\mathcal{W}^s$ of initial states for system (2.11)-(2.16) is controllable in time T then there exists a sequence $(v_k)_{k\in\mathbb{Z}_*}$ biorthogonal to $\left(\varkappa_{|k|}e^{i\lambda_k t}\right)_{k\in\mathbb{Z}_*}$ in $L^2(0,T)$, which satisfies*

$$||v_k||_{L^2(0,T)} \le C\lambda_k^{s-1}, \qquad k\in\mathbb{Z}_*,$$

where C is a positive constant independent of k.

Proof. If the subspace $\mathcal{W}^s$ is controllable in time T, there exists a constant $C>0$ such that

$$\int_0^T |\partial_n\phi^1(t,\mathbf{v}_1)|^2 dt \ge C||(\bar{\phi}_0,\bar{\phi}_1)||^2_{V^{1-s}\times V^{-s}}.$$

Then, in view of Corollary 3.7, for every $(\bar{u}_0, \bar{u}_1)$ there exists $h \in L^2(0, T)$ such that

$$\|h\|_{L^2(0,T)} \leq C\|(\bar{u}_0, \bar{u}_1)\|_{W^s}.$$

Thus, the functions g_m, h_m constructed in Proposition 3.24 satisfy

$$\|g_m\|_{L^2(0,T)} \leq C\lambda_m^s, \qquad \|h_m\|_{L^2(0,T)} \leq C\lambda_m^{s-1}.$$

Then, from (3.41) it holds

$$\|v_m\|_{L^2(0,T)} \leq C\lambda_m^{s-1}.$$

Now it suffices to recall that the sequence $(v_m)_{m \in \mathbb{Z}_*}$ is biorthogonal to $\left(\varkappa_{|k|} e^{i\lambda_k t}\right)_{k \in \mathbb{Z}_*}$.

Remark 3.26. If we perform the change of variable $t \to t - T/2$ we obtain that the assertions of Propositions 3.24 and 3.25 remain true if we replace the space $L^2(0, T)$ by $L^2(-T/2, T/2)$.

Remark 3.27. The numbers $\varkappa_k = \partial_n \theta_k^1(\mathbf{v}_1)$ have a direct incidence in the spectral controllability of system (2.11)-(2.16). If $\varkappa_k = 0$ for some k then, from (3.36), (3.37) it follows that the initial state $(\bar{u}_0, \bar{u}_1)$ is controllable only if $u_{0,k} = u_{1,k} = 0$, that is, if $\bar{u}_0$ and $\bar{u}_1$ are orthogonal to $\bar{\theta}_k$. In this case, the space of controllable initial states is not dense in $H \times V'$. Consequently, the condition $\varkappa_k \neq 0$ for every $k \in \mathbb{N}$ is necessary for approximate controllability (and in particular for the spectral one) of system (2.11)-(2.16).

For the sequence $(|\varkappa_k|)$ an upper bound may be easily obtained. If we consider the solutions

$$\bar{\phi}(t, x) = \cos \lambda_k t \, \bar{\theta}_k(x), \quad k \in \mathbb{N},$$

of the homogeneous system (2.17)-(2.21) and apply inequality (2.29) it holds

$$|\varkappa_k|^2 \int_0^T |\cos \lambda_k t|^2 \, dt = \int_0^T \left|\phi_{k,x}^1(t, \mathbf{v}_1)\right|^2 dt \leq C\mathbf{E}_{\bar{\phi}} = C\lambda_k^2.$$

In an analogous way, taking $\bar{\phi}(t, x) = \sin \lambda_k t \, \bar{\theta}_k(x)$ we have

$$|\varkappa_k|^2 \int_0^T |\sin \lambda_k t|^2 \, dt = \int_0^T \left|\phi_{k,x}^1(t, \mathbf{v}_1)\right|^2 dt \leq C\mathbf{E}_{\bar{\phi}} = C\lambda_k^2.$$

From these two inequalities we see that the sequence $\varkappa_k$ satisfies

$$|\varkappa_k| \leq C\lambda_k, \quad k \in \mathbb{N}. \tag{3.42}$$

3.4 Riesz Bases and Ingham-Type Inequalities

In this section we describe the technique of proof of observability inequalities based on various variants of the so called Ingham inequality which provides a Riesz basis of subspaces of $L^2(0, T)$ generated by finite linear combinations of complex exponentials $(e^{i\lambda_n t})$. In particular, we shall use a generalization of that classical result recently proved by Baiocchi, Komornik and Loreti in [17] and Avdonin and Moran in [11]. One of the main consequences of the developments in this section is Theorem 3.34 which guarantees that, if we prove an Ingham-type inequality for the sequence (λ_n), then a similar inequality holds for the sequence (λ_n^s) with $s > 1$. This is a useful tool to derive controllability properties for Schrödinger and beam-like equations in networks from those obtained for string equations.

3.4.1 Riesz Bases

In general, if $\mathbf{H}$ is a separable Hilbert space, the sequence $(\mathbf{a}_n) \subset \mathbf{H}$ is called *Riesz basis of the closure of its linear span* if there exist constants $c_1, c_2 > 0$ such that the following inequality is verified

$$c_1 \|\bar{\gamma}\|_{l^2} \leq \|\sum_{n \in \mathbb{Z}} \gamma_n \mathbf{a}_n\|_H \leq c_2 \|\bar{\gamma}\|_{l^2},$$

for every finite sequence of complex numbers such that $\bar{\gamma} = (\gamma_n)$. In particular, if the sequence $(\mathbf{a}_n)$ is complete in $\mathbf{H}$ it is called *Riesz basis*[4] *of* $\mathbf{H}$.

Thus, the fact that the sequence $(e^{i\lambda_n t})$ forms a Riesz basis of $L^2(0, T)$ is very useful to prove observability inequalities (3.20).

Let us observe that, essentially, the technique derived from the use of Riesz bases coincides with the method of moments, since a theorem due to Bari [19] asserts that the inequality

$$c_1 \|\bar{\gamma}\|_{l^2} \leq \|\sum_{n \in \mathbb{Z}} \gamma_n \mathbf{a}_n\|_H$$

is equivalent to the fact that the problem of moments (3.24) has a solution for any $(m_n) \in l^2$.

3.4.2 Generalized Ingham Theorems

An important theorem due to Ingham [54] asserts that the sequence $(e^{i\lambda_n t})$ forms a Riesz basis of the closure of its linear span in $L^2(0, T)$ if the sequence (λ_n) satisfies the separation condition

---

[4] An equivalent definition is that $(\mathbf{a}_n)$ is the image of an orthonormal basis of $\mathbf{H}$ by a continuous bijection. In [116] the reader may find more information on this topic.

$$\lambda_{n+1} - \lambda_n \geq \gamma > 0, \tag{3.43}$$

and $\gamma > 2\pi/T$.

A stronger version of this result was given by Beurling in [20]: if the sequence (λ_n) satisfies the condition (3.43), then $(e^{i\lambda_n t})$ forms a Riesz basis in the closure of its linear span in $L^2(0, T)$ for every T satisfying

$$T > 2\pi D^+(\lambda_n),$$

where $D^+(\lambda_n)$ is the upper density of the sequence (λ_n):

$$D^+(\lambda_n) := \lim_{r \to \infty} \frac{n^+(r, (\lambda_n))}{r},$$

with $n^+(r, (\lambda_n))$ being the maximum number of elements of (λ_n) contained in an interval of length r.

The inequality corresponding to this assertion

$$C_1 \|\bar{c}\|_{l^2}^2 \leq \int_0^T \left| \sum_{n \in \mathbb{Z}} c_n e^{i\lambda_n t} \right|^2 dt \leq C_2 \|\bar{c}\|_{l^2}^2, \tag{I}$$

is known as *Ingham inequality*. This inequality has been an extremely useful tool in the study of control problems.

In many specific problems, however, the separation condition (3.43) is not verified. This is the case, for example, of the networks of strings (see Proposition 4.23). Consequently, a lot of work has been devoted to obtaining various variants and weakened versions of (I) of the type

$$\int_0^T \left| \sum_{n \in \mathbb{Z}} c_n e^{i\lambda_n t} \right|^2 dt \geq C \|\mathbf{B}\bar{c}\|_{l^2}^2, \tag{I_B}$$

where $\mathbf{B} : l^2 \to l^2$ is a continuous operator, usually with a simple structure, when the sequence (λ_n) does not satisfy the gap condition (3.43). We refer to the works [27], [29], [57], [56], [11], [15], [16], [17] for further information.

Let us observe that an inequality of type (I_B) guarantees that the problem of moments has a solution for every $\bar{c}$ in the image of the adjoint of $\mathbf{B}$. This subspace is necessarily smaller than l^2 if the sequence (λ_n) does not satisfy the gap condition (3.43). Indeed, otherwise, $\mathbf{B}$ would have a bounded inverse, and this would lead to the Ingham inequality (I), which is not true in the case of lack of gap.

The most complete result in this direction was independently obtained by Baiocchi, Komornik and Loreti in [15], [16], [17] and Avdonin and Moran in [11]. In their papers an inequality of the form (I_B) is proved for increasing sequences of real numbers (λ_n) with the following generalized separation property:

There exist $\delta > 0$ and a natural number M such that

$$\lambda_{n+M} - \lambda_n \geq M\delta \tag{3.44}$$

for every $n \in \mathbb{Z}$.

This means that there may be at most M consecutive elements of the sequence (λ_n) that are close to each other, while, in a larger number, there must be some gap.

Remark 3.28. The separation property (3.44) may be described in an equivalent way in terms of the upper density of the sequence (λ_n). It turns out that, if $D^+(\lambda_n)$ is finite and $T > 2\pi D^+(\lambda_n)$, then there exist $\delta > 2\pi/T$ and $M \in \mathbb{N}$ such that (λ_n) satisfies the separation condition (3.44). The details of the proof may be found in [17].

In order to state the main result of the papers mentioned above and to describe how the corresponding operator $\mathbf{B}$ is constructed, we need some preliminary elements.

Let us fix a sequence (λ_n) satisfying the separation condition (3.44). We will say that two integer numbers n, m are equivalent if

$$|\lambda_n - \lambda_m| < |n - m|\,\delta.$$

This is an equivalence relation in $\mathbb{Z}$. Let us denote by Λ_k, $k \in \mathbb{Z}$, the equivalence classes of $\mathbb{Z}$ with respect to this relation. Obviously, each Λ_k is formed by consecutive numbers and contains $d(k) \leq M$ elements. We denote by $n(k)$ the smallest element of Λ_k. Besides, we assume that the numbering of the classes has been chosen so that $n(k+1) - 1 \in \Lambda_k$, that is, $n(k+1) = n(k) + d(k)$.

For every $m \in \mathbb{N}$ we pick k such that $m \in \Lambda_k$ and define the function

$$f_m(t) = \sum_{j=n(k)}^{m} \frac{e^{i\lambda_j t}}{\pi_{j,m}},$$

where $\pi_{j,m}$ is the product of all the differences $\lambda_m - \lambda_j$ with $n(k) \leq j < m$ if $n(k) < m$ and $\pi_{n(k),n(k)} = 1$. These functions are called *divided differences of the family* $(e^{i\lambda_n t})$ (see [55], p. 246 or [111]).

Theorem 3.29 (Baiocchi *et al.* [17], Avdonin-Moran [11]). *For all the values of $\delta > 0$, $M \in \mathbb{N}$ and $T > 2\pi/\delta$ if (λ_n) satisfies the separation condition (3.44), then the sequence (f_n) forms a Riesz basis in the closure of its linear span in $L^2(0,T)$.*

The following result is also proved in [11]. It allows clarifying what happens when the value of T is not sufficiently large.

Theorem 3.30. *If the sequence (λ_n) satisfies the gap condition (3.44) and $T < 2\pi D^+$ then there exists a proper subsequence $(\hat{n}) \subset \mathbb{Z}$ such that $(f_{\hat{n}})$ forms a Riesz basis in $L^2(0,T)$.*

As a consequence, after applying Theorem III.3.10(e) in [9], it holds

Corollary 3.31. *If the sequence* (λ_n) *satisfies the condition* $D^+(\lambda_n) < \infty$ *then, for every* $T < 2\pi D^+(\lambda_n)$, *there exist complex numbers* c_n, *not all of then equal to zero, such that*

$$\sum_{n \in \mathbb{Z}} |c_n|^2 < \infty, \qquad \sum_{n \in \mathbb{Z}} c_n e^{i\lambda_n t} = 0 \quad a. \ e. \ (0, T).$$

Let us write the positive result of Theorem 3.29 as an inequality of type $(\mathbf{I_B})$.

Let $\bar{\mathbf{e}}^n$, $n \in \mathbb{Z}$, be the canonical basis of l^2 and consider the subspaces

$$\mathbf{L}_k = \mathrm{span}_{n \in \Lambda_k} (\bar{\mathbf{e}}^n).$$

Each subspace $\mathbf{L}_k$ has finite dimension $d(k) \leq M$. Then, l^2 is decomposed as

$$l^2 = \bigoplus_{k \in \mathbb{Z}} \mathbf{L}_k.$$

Let $m \in \mathbb{N}$. For every $\bar{h} = (h_1, ..., h_m) \in \mathbb{R}^m$, we define the operators $\mathbf{A}_m(\bar{h}) : \mathbb{R}^m \to \mathbb{R}^m$ by $\mathbf{A}_m(\bar{h})\bar{x} = A_m(\bar{h})\bar{x}$, where $A_m(\bar{h})$ is the matrix with components

$$\mathbf{A}_{m,ij}(\bar{h}) = \begin{cases} \prod_{k=1}^{\prime j} (h_i - h_k)^{-1} & \text{if } i \leq j, \\ 1 & \text{if } i = j = 1, \\ 0 & \text{if } i > j, \end{cases} \tag{3.45}$$

where the symbol $'$ in the product sign indicates that the factor corresponding to $k = i$ has been excluded.

These matrices are invertible if all the numbers h_j are pairwise distinct. Now we take

$$\mathbf{B}_k = \left(\mathbf{A}_{d(k)}(\lambda_{n(k)}, ..., \lambda_{n(k)+d(k)-1}) \right)^{-1}.$$

Finally, the operator $\mathbf{B}$ is defined for $\bar{\mathbf{v}} = \sum_{k \in \mathbb{Z}} \bar{\mathbf{v}}_k$ as

$$\mathbf{B}\bar{\mathbf{v}} = \sum_{k \in \mathbb{Z}} \mathbf{B}_k \bar{\mathbf{v}}_k,$$

where $\bar{\mathbf{v}}_k$ is the projection of $\bar{\mathbf{v}}$ over $\mathbf{L}_k$. This is the operator appearing in inequality $(\mathbf{I_B})$ corresponding to the assertion of Theorem 3.29.

Theorem 3.32. *For all* $\delta > 0$, $M \in \mathbb{N}$ *and* $T > 2\pi/\delta$, *there exist constants* $C_1, C_2 > 0$ *such that, if the sequence* (λ_n) *satisfies the separation condition (3.44) then*

$$C_1 \|\mathbf{B}\bar{c}\|_{l^2}^2 \geq \int_0^T \left| \sum_{n \in \mathbb{Z}} c_n e^{i\lambda_n t} \right|^2 dt \geq C_2 \|\mathbf{B}\bar{c}\|_{l^2}^2, \tag{$\mathbf{I_B}$}$$

for every finite sequence $\bar{c}$.

We should remark that the operator $\mathbf{B}$ has a structure that makes it easy to obtain information from inequality $(I_{\mathbf{B}})$. According to its definition we have

$$\|\mathbf{B}\bar{c}\|_{l^2}^2 = \sum_{k \in \mathbb{Z}} \|\mathbf{B}_k \bar{c}_k\|_{l^2}^2 = \sum_{k \in \mathbb{Z}} \left\| \left(\mathbf{A}_{d(k)}(\lambda_{n(k)}, ..., \lambda_{n(k)+d(k)-1}) \right)^{-1} \bar{c}_k \right\|_{l^2}^2$$

$$\geq \sum_{k \in \mathbb{Z}} \gamma_k^2 \|\bar{c}_k\|_{l^2}^2 \,,$$

where

$$\gamma_k = \left\| \left(\mathbf{A}_{d(k)}(\lambda_{n(k)}, ..., \lambda_{n(k)+d(k)-1}) \right) \right\|^{-1}.$$

Taking into account that

$$\|\bar{c}_k\|_{l^2}^2 = \sum_{n=n(k)}^{n(k)+d(k)-1} |c_n|^2 \,,$$

from inequality (3.44) it holds

$$\int_0^T \left| \sum_{n \in \mathbb{Z}} c_n e^{i\lambda_n t} \right|^2 dt \geq C_2 \sum_{k \in \mathbb{Z}} \gamma_k^2 \sum_{n=n(k)}^{n(k)+d(k)-1} |c_n|^2.$$

This simply says that it is sufficient to choose weights γ_k^2 in the coefficients corresponding to $n \in \Lambda_k$. Thus, we have obtained

Corollary 3.33. *If the strictly increasing sequence (λ_n) satisfies the gap condition (3.44) or, equivalently, $D^+(\lambda_n) < \infty$ then, for every $T > 2\pi D^+(\lambda_n)$ there exist positive numbers γ_n, such that*

$$\int_0^T \left| \sum_{n \in \mathbb{Z}} c_n e^{i\lambda_n t} \right|^2 dt \geq \sum_{n \in \mathbb{Z}} \gamma_n^2 |c_n|^2 \,,$$

for every finite sequence (c_n).

It is possible to get more precise estimates on $\left\| \mathbf{A}_{d(k)}^{-1} \bar{c}_k \right\|_{l^2}^2$ leading to inequalities with weights, which vary inside of each block Λ_k. But this depends on the particular structure of the operator $\mathbf{A}_m$, and of course, on the sequence (λ_n).

3.4.3 A New Inequality

The following result turns out to be very useful for the identification of subspaces of controllable initial states for the Schrödinger and beam equations once we know subspaces of controllable initial states for system (2.11)-(2.16). This result will be used in Chapter 8.

Theorem 3.34. *Let (λ_n) be an increasing sequence of positive numbers with upper density $D^+(\lambda_n) < \infty$. Assume that there exist constants $C > 0$ and $\alpha < 0$ such that the inequality*

$$\int_0^T \left| \sum_{n \in \mathbb{Z}} c_n e^{i\lambda_n t} \right|^2 dt \geq C \sum_{n \in \mathbb{Z}} \lambda_n^{2\alpha} c_n^2, \tag{3.46}$$

is verified for every finite sequence $\bar{c}$. Then, for all $\tau > 0$ and $s > 1$, there exists a constant $C_1 > 0$ such that

$$\int_0^\tau \left| \sum_{n \in \mathbb{Z}} c_n e^{i\lambda_n^s t} \right|^2 dt \geq C_1 \sum_{n \in \mathbb{Z}} \lambda_n^{2\alpha} c_n^2,$$

for every finite sequence $\bar{c}$.

The proof of this assertion is based on Theorem 3.32. We will need the following technical results.

Proposition 3.35. *Let $K : \mathbb{R}^m \to \mathbb{R}^m$ be a linear operator defined by the matrix $A = (a_{ij})$. If $\|A\|$ is the norm of A considered as a linear operator from $l^2(\mathbb{R}^m)$ to $l^2(\mathbb{R}^m)$ then*

$$\max_{i,j=1,\ldots,m} |a_{ij}| \leq \|A\| \leq \sqrt{m} \max_{i,j=1,\ldots,m} |a_{ij}|.$$

This fact is easily proved with the aid of Schur's Lemma or directly using the Cauchy-Schwarz inequality (see, e.g., [52], Theorem 3.4.7).

Proposition 3.36. *Let $\mathbf{A}_m(\bar{h}) : \mathbb{R}^m \to \mathbb{R}^m$ be defined by (3.45) and assume that $1 < h_1 \leq h_2 \leq \cdots \leq h_m$. If $\bar{h}^s = (h_1^s, \ldots, h_m^s)$ ($\bar{h}^s$ is formed by the s-powers of the components of $\bar{h}$). Then, for every $s > 1$,*

$$\left\| \mathbf{A}_m(\bar{h}^s) \right\| \leq \sqrt{m} \left\| \mathbf{A}_m(\bar{h}) \right\|. \tag{3.47}$$

Proof. Let us observe that

$$\left| h_i^s - h_j^s \right| \geq s \left| h_i - h_j \right| h_1^{s-1}$$

and thus, from the definition of $\mathbf{A}_{m,ij}(\bar{h}^s)$, we obtain

$$\left| \mathbf{A}_{m,ij}(\bar{h}^s) \right| \leq \left| \mathbf{A}_{m,ij}(\bar{h}) \right| h_1^{(s-1)(j-1)} s^{(j-1)}.$$

Now, using Proposition 3.35,

$$
\begin{aligned}
\left\| \mathbf{A}_m(\bar{h}^s) \right\| &\leq \sqrt{m} \max_{i,j=1,\ldots,m} \left| \mathbf{A}_{m,ij}(\bar{h}^s) \right| \\
&\leq \sqrt{m} \max_{i,j=1,\ldots,m} \left| \mathbf{A}_{m,ij}(\bar{h}) \right| h_1^{(s-1)(j-1)} s^{(j-1)} \\
&\leq \sqrt{m} \max_{i,j=1,\ldots,m} \left| \mathbf{A}_{m,ij}(\bar{h}) \right| \max_{j=1,\ldots,m} h_1^{(s-1)(1-j)} s^{(1-j)}.
\end{aligned}
$$

Taking into account that $h_1 > 1$ it follows

$$\max_{j=1,\ldots,m} h_1^{(s-1)(1-j)} s^{(1-j)} = 1$$

and then

$$\left\| \mathbf{A}_m(\bar{h}^s) \right\| \leq \sqrt{m} \max_{i,j=1,\ldots,m} \left| \mathbf{A}_{m,ij}(\bar{h}) \right|.$$

Applying once again Proposition 3.35, inequality (3.47) is obtained.

Proof (Proof of Theorem 3.34). Let us choose $\delta < 1/D^+$ and $T > 2\pi/\delta$. According to Theorem 3.32, there exists a constant $C_1 > 0$ such that

$$C_1 \|\mathbf{B}\bar{c}\|_{l^2}^2 \geq \int_0^T \left| \sum_{n \in \mathbb{Z}} c_n e^{i\lambda_n t} \right|^2.$$

In account of the hypothesis (3.46) of the lemma and the fact

$$\|\mathbf{B}\bar{c}\|_{l^2}^2 = \sum_{k \in \mathbb{Z}} \left\| \left(\mathbf{A}_{d(k)}(\lambda_{n(k)}, \ldots, \lambda_{n(k)+d(k)-1}) \right)^{-1} \bar{c}_k \right\|_{l^2}^2,$$

we obtain

$$\left\| \left(\mathbf{A}_{d(k)}(\lambda_{n(k)}, \ldots, \lambda_{n(k)+d(k)-1}) \right)^{-1} \bar{c}_k \right\|_{l^2}^2 \geq C \sum_{n \in \Lambda_k} \lambda_n^{2\alpha} c_n^2,$$

for all $\bar{c}_k \in \mathbb{R}^{d(k)}$. Since the sequence (λ_n) is increasing, this inequality implies

$$\left\| \left(\mathbf{A}_{d(k)}(\lambda_{n(k)}, \ldots, \lambda_{n(k)+d(k)-1}) \right)^{-1} \bar{c}_k \right\|_{l^2}^2 \geq C \lambda_{n(k)+d(k)-1}^{2\alpha} \sum_{n \in \Lambda_k} c_n^2.$$

Thus, we can conclude that

$$\left\| \mathbf{A}_{d(k)}(\lambda_{n(k)}, \ldots, \lambda_{n(k)+d(k)-1}) \right\| \leq C \lambda_{n(k)+d(k)-1}^{-\alpha}.$$

Now, Proposition 3.36 allows us to ensure that

$$\left\| \mathbf{A}_{d(k)}(\lambda_{n(k)}^s, \ldots, \lambda_{n(k)+d(k)-1}^s) \right\| \leq C \lambda_{n(k)+d(k)-1}^{-\alpha}. \tag{3.48}$$

Let $\tau > 0$ and choose n_0 such that $\delta' := \delta s \lambda_{n_0}^{s-1} > 2\pi/\tau$. Then, for every $n \geq n_0$ it holds

$$\lambda_{n+M}^s - \lambda_n^s > s \left(\lambda_{n+M} - \lambda_n \right) \lambda_n^{s-1} \geq M \delta s \lambda_n^{s-1} = M\delta'.$$

In particular, every set of the partition (Λ_k^s) of $\mathbb{Z}$, defined for the sequence (λ_n^s) for the value δ', which contains and element $n \geq n_0$, will be contained in one of the sets Λ_k.

Once again from Theorem 3.32 we obtain that for every finite sequence $\bar{c}$,

$$\int_0^T \left| \sum_{n \in \mathbb{Z}} c_n e^{i\lambda_n^s t} \right|^2 dt \geq C_2 \left\| \mathbf{B}^s \bar{c} \right\|_{l^2}^2 . \tag{3.49}$$

Here, the operator $\mathbf{B}^s$ corresponds to the sequence (λ_n^s) and to δ', that is, to the partition (Λ_k^s).

It is possible to prove (see Lemma 3.1 in [17]) that, if $\mathbf{B}_\delta$ and $\mathbf{B}_{\delta'}$ are the operators defined by (3.45) for the partitions generated by δ and δ', respectively, then there exist constants $d_1, d_2 > 0$, depending only on δ and δ', such that

$$d_1 \left\| \mathbf{B}_\delta \bar{c} \right\|_{l^2}^2 \leq \left\| \mathbf{B}_{\delta'} \bar{c} \right\|_{l^2}^2 \leq d_2 \left\| \mathbf{B}_\delta \bar{c} \right\|_{l^2}^2 ,$$

for every finite sequence $\bar{c}$.

Thus, we may assume that the operator $\mathbf{B}^s$ has been constructed for the partition (Λ_k).

Now it suffices to note that

$$\left\| \mathbf{B}^s \bar{c} \right\|_{l^2}^2 \geq \sum_{k \in \mathbb{Z}} \left(\left\| \mathbf{A}_{d(k)}(\lambda_{n(k)}, ..., \lambda_{n(k)+d(k)-1}) \right\|_{l^2}^{-1} \right)^2 \left\| \bar{c}_k \right\|_{l^2}^2$$

and to use the inequality (3.48) to conclude that

$$\int_0^T \left| \sum_{n \in \mathbb{Z}} c_n e^{i\lambda_n^s t} \right|^2 dt \geq C_2 \sum_{k \in \mathbb{Z}} \lambda_{n(k)+d(k)-1}^{2\alpha} \left\| \bar{c}_k \right\|_{l^2}^2 .$$

Finally, let us observe that there exists a constant $C > 0$ such that for every n satisfying $n(k) \leq n \leq n(k) + d(k) - 1$, it holds

$$\lambda_{n(k)+d(k)-1} \leq C\lambda_n.$$

This concludes the proof.

4

The Three String Network

This chapter is devoted to study the control problem for the simplest non trivial network of strings that cannot be reduced to a single string: *the three string network*. Most of the results presented here will be generalized later in Chapter 5 to the case of general networks supported by tree-shaped graphs. However, the generality of the problem in that case involves complex notations. It is therefore convenient to first address the simple case of the three string network, for which the main ideas involved in our analysis, that will allow us to address the case of general networks, can be described more transparently.

We first consider the case when two of the three external nodes of the network are controlled. In this case, standard methods based on the d'Alembert formula and energy arguments allow showing that the observability and controllability properties hold in the optimal energy space. We then address the case when a single control acts on one exterior node. In this case the problem is much more complex since the space in which observability and controllability hold depend on the irrationallity properties of the ratios of the lengths of the strings entering in the network. The methods to analyze it also vary significantly and are based on results from Number Theory.

4.1 The Three String Network with Two Controlled Nodes

4.1.1 Equations of Motion of the Network

Let T, ℓ_0, ℓ_1, ℓ_2 be positive numbers. We consider the following non-homogeneous system

$$\begin{cases} u_{tt}^i - u_{xx}^i = 0 & \text{in } \mathbb{R}\times[0,\ell_i],\ i=0,1,2, \\ u^0(t,0) = u^1(t,0) = u^2(t,0) & t \in \mathbb{R}, \\ u_x^0(t,0) + u_x^1(t,0) + u_x^2(t,0) = 0 & t \in \mathbb{R}, \\ u^i(t,\ell_i) = v^i(t), \quad u^2(t,\ell_2) = 0 & t \in \mathbb{R}, \qquad i=0,1, \\ u^i(0,x) = u_0^i(x), \quad u_t^i(0,x) = u_1^i(x)\ x \in [0,\ell_i], \quad i=0,1,2 \end{cases} \qquad (4.1)$$

which models the vibrations of a network formed by three elastic strings $\mathbf{e}_0, \mathbf{e}_1, \mathbf{e}_2$ with lengths ℓ_0, ℓ_1, ℓ_2 coupled at one of their extremes. The functions $u^i = u^i(t,x) : [0,\ell_i] \to \mathbb{R}$, $i = 0,1,2$, represent the transversal displacements of the strings. On the free nodes of the strings $\mathbf{e}_0$ and $\mathbf{e}_1$ some external controls v^0 and v^1 act regulating their motion.

Let us observe that in (4.1), the parametrization of the strings has been chosen so that $x = 0$ corresponds to the common node, while $x = \ell_i$ correspond to the exterior nodes of the strings $\mathbf{e}_i$, $i = 1,2$.

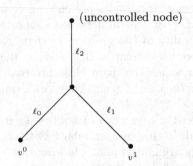

Fig. 4.1. The three string network with two controlled nodes

Let $T > 0$. According to the general results described in Chapter 2, the homogeneous system resulting in the absence of control in (4.1) ($v^0 = v^1 = 0$)

$$\begin{cases} \phi_{tt}^i - \phi_{xx}^i = 0 & \text{in } \mathbb{R}\times[0,\ell_i],\ i=0,1,2, \\ \phi^0(t,0) = \phi^1(t,0) = \phi^2(t,0) & t \in \mathbb{R}, \\ \phi_x^0(t,0) + \phi_x^1(t,0) + \phi_x^2(t,0) = 0 & t \in \mathbb{R}, \\ \phi^i(t,\ell_i) = 0 & t \in \mathbb{R}, \qquad i=0,1,2, \\ \phi^i(0,x) = \phi_0^i(x), \quad \phi_t^i(0,x) = \phi_1^i(x)\ x \in [0,\ell_i], \quad i=0,1,2, \end{cases} \qquad (4.2)$$

has a unique solution $\bar{\phi}$ with initial state $(\bar{\phi}_0, \bar{\phi}_1) \in V \times H$ satisfying

$$\bar{\phi} \in C([0,T]:V)\bigcap C^1([0,T]:H). \qquad (4.3)$$

Recall that the spaces V and H are those defined in Chapter 2, Section 2.2, that is

$$V = \left\{ \bar{\phi} \in \prod_{i=0}^{2} H^1(0, \ell_i) : \phi_0(0) = \phi_1(0) = \phi_2(0), \ \phi_i(\ell_i) = 0, \ i = 0, 1, 2 \right\},$$

$$H = \prod_{i=0}^{2} L^2(0, \ell_i).$$

The solution $\bar{\phi}$ of (4.2) is expressed in terms of the Fourier coefficients $(\phi_{0,n})$, $(\phi_{1,n})$ of the initial data in the orthonormal basis $(\bar{\theta}_n)$ formed by the eigenfunctions of the elliptic operator $-\Delta_G$ associated to the star-like network under consideration, by the formula

$$\bar{\phi}(t, x) = \sum_{n \in \mathbb{N}} \left(\phi_{0,n} \cos \lambda_n t + \frac{\phi_{1,n}}{\lambda_n} \sin \lambda_n t \right) \bar{\theta}_n(x). \tag{4.4}$$

The energy of $\bar{\phi}$, defined as the sum of the energies of the solutions on the strings, is constant in time and, according to the relation (2.25), it may be expressed as

$$\mathbf{E}_{\bar{\phi}} = \sum_{n \in \mathbb{N}} (\mu_n \phi_{0,n}^2 + \phi_{1,n}^2). \tag{4.5}$$

On the other hand, for the non-homogeneous system (4.1), for every $v^0, v^1 \in L^2(0, T)$, there exists a unique solution, defined by transposition that satisfies

$$\bar{u} \in C([0, T] : H) \bigcap C^1([0, T] : V').$$

Here and in the sequel V' denotes the dual of V.

4.1.2 The Control Problem

The control problem in time T for system (4.1) consists in *characterizing the initial states $(\bar{u}_0, \bar{u}_1) \in H \times V'$ of the network for which there exist controls $v^0, v^1 \in L^2(0, T)$ such that the corresponding solution of (4.1) satisfies*

$$\bar{u}(T) = \bar{u}_t(T) = \bar{0}.$$

The control of a three string network from two exterior nodes satisfies the hypotheses of Theorem 2.7. In this case it holds

Theorem 4.1. *System (4.1) is exactly controllable in time*

$$T^* = 2(\ell_2 + \max\{\ell_0, \ell_1\}).$$

Proof. Let us assume that $\ell_0 \geq \ell_1$, such that $T^* = 2(\ell_0 + \ell_2)$. In view of Proposition 2.5, the initial state $(\bar{u}_0, \bar{u}_1) \in H \times V'$ is controllable in time T with controls $v^0, v^1 \in L^2(0, T)$ if, and only if,

$$\int_0^{T^*} \phi_x^0(t, \ell_0) v^0(t) dt + \int_0^{T^*} \phi_x^1(t, \ell_1) v^1(t) dt = -\langle \bar{u}_0, \bar{\phi}_1 \rangle_H + \langle \bar{u}_1, \bar{\phi}_0 \rangle_{V' \times V},$$

for every solution $\bar{\phi}$ of system (4.2) with initial data $(\bar{\phi}_0, \bar{\phi}_1) \in Z \times Z$. Corollary 3.16 of Theorem 3.4 allows us to ensure that system (4.1) is exactly controllable in time T if, and only if, there exists a constant $C > 0$ such that

$$\int_0^{T^*} |\phi_x^0(t, \ell_0)|^2 dt + \int_0^{T^*} |\phi_x^1(t, \ell_1)|^2 dt \geq C \mathbf{E}_{\bar{\phi}}, \qquad (4.6)$$

for every solution $\bar{\phi}$ of the homogeneous system (4.2) with initial state in $Z \times Z$.

In order to prove inequality (4.6), in view of the property of conservation of energy, it suffices to find $\hat{t} \in \mathbb{R}$ such that

$$\int_0^{2(\ell_0 + \ell_2)} \left(|\phi_x^0(t, \ell_0)|^2 + |\phi_x^1(t, \ell_1)|^2 \right) dt \geq C \mathbf{E}_{\phi^i}(\hat{t}), \qquad i = 0, 1, 2. \quad (4.7)$$

Thanks to Proposition 3.1 estimate (4.7) holds immediately for $i = 0, 1$ (that is, for the components of the solution corresponding to the controlled strings) if $\hat{t} \in [\ell_0, 2\ell_2 + \ell_0]$.

It remains to recover an estimate of the energy of the string corresponding to $i = 2$. The idea is very simple: the d'Alembert formula allows proving the identities

$$\phi_x^0(t, 0) = \ell_0^+ \phi_x^0(t, \ell_0), \qquad \phi_t^0(t, 0) = \ell_0^- \phi_x^0(t, \ell_0), \qquad (4.8)$$

$$\phi_x^1(t, 0) = \ell_1^+ \phi_x^1(t, \ell_1), \qquad \phi_t^1(t, 0) = \ell_1^- \phi_x^1(t, \ell_1). \qquad (4.9)$$

Moreover, in account of the transmission conditions in the common node, we have

$$\phi_t^2(t, 0) = \phi_t^1(t, 0) = \ell_1^- \phi_x^1(t, \ell_1),$$

$$\phi_x^2(t, 0) = - \left(\phi_x^0(t, 0) + \phi_x^1(t, 0) \right) = \left(\ell_0^+ \phi_x^0(t, \ell_0) + \ell_1^+ \phi_x^1(t, \ell_1) \right).$$

Then, according to Proposition 3.1,

$$\mathbf{E}_{\phi^2}(\hat{t}) \leq \int_{\hat{t} - \ell_2}^{\hat{t} + \ell_2} \left(|\phi_t^2(t, 0)|^2 + |\phi_x^2(t, 0)|^2 \right) dt$$

$$= \int_{\hat{t} - \ell_2}^{\hat{t} + \ell_2} \left(|\ell_1^- \phi_x^1(t, \ell_1)|^2 + |\ell_0^+ \phi_x^0(t, \ell_0) + \ell_1^+ \phi_x^1(t, \ell_1)|^2 \right) dt.$$

From this inequality and applying Proposition 3.3 we obtain

$$C\mathbf{E}_{\phi^2}(\hat{t}) \leq \int_{\hat{t}-\ell_2-\ell_1}^{\hat{t}+\ell_2+\ell_1} \left|\phi_x^1(t,\ell_1)\right|^2 dt + \int_{\hat{t}-\ell_0-\ell_2}^{\hat{t}+\ell_0+\ell_2} \left|\phi_x^0(t,\ell_0)\right|^2 dt,$$

and thus, choosing $\hat{t} = \ell_0 + \ell_2$, all the inequalities (4.7) are verified.

Remark 4.2. It is clear that the same procedure would work in the case of a general tree-shaped network controlled from all of its exterior nodes, except one: it suffices to apply an induction argument.

The application of the d'Alembert formula and Proposition 3.1 allow to estimate the norms

$$\int_{\beta+\ell}^{\alpha-\ell} |\phi_x(t,\ell)|^2 dt, \qquad \int_{\beta+\ell}^{\alpha-\ell} |\phi_t(t,\ell)|^2 dt$$

of the traces ϕ_x and ϕ_t in the extreme $x = \ell$ from the norms

$$\int_{\beta}^{\alpha} |\phi_x(t,0)|^2 dt, \qquad \int_{\beta}^{\alpha} |\phi_t(t,0)|^2 dt$$

of the traces ϕ_x and ϕ_t in the extreme $x = 0$, provided ϕ solves the wave equation in the space interval $0 < x < \ell$. Thus, in the case of a general tree-shaped network, we start from the controlled nodes (for which $\phi = 0$ and consequently $\phi_t = 0$ and $\phi_x \in L^2(0,T)$) and apply this argument up to the first interior node. Because of the tree-like structure, this provides estimates of the traces ϕ_x and ϕ_t of all the components that are coupled in that node, except for one of them. The two coupling conditions on that node allow then to obtain estimates of the traces of ϕ_t and ϕ_x for the remaining string corresponding to that node. This argument allows getting estimates on all the interior nodes that can be joined to the controlled exterior ones by one single string: the first layer of interior nodes. One can iterate this argument to get control of all the strings of the network. This yields the result by G. Schmidt [108] in Section 2.4.

4.2 A Simpler Problem: Simultaneous Control of Two Strings

The simultaneous control problem of two strings $\mathbf{e}_1$ and $\mathbf{e}_2$ of lengths ℓ_1 and ℓ_2 is similar to the previous one. It was implicitly studied in [57]. Later, in [110] and [14] an essentially complete solution was obtained. The results of [110] are based on a generalization of the Ingham inequality proved in [57]. This technique, however, allowed only to guarantee the controllability of the system in a time larger than the optimal one. In [14] the method of moments was used; this method provided the optimal control time. Here we describe a different method, based on elementary arguments, which in addition provides more information than the previously mentioned techniques.

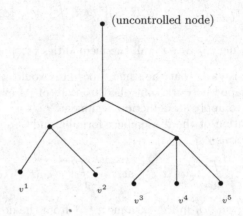

Fig. 4.2. Tree-shaped network with one uncontrolled node

The system corresponding to the simultaneous control of two strings is

$$
\begin{cases}
u_{tt}^i - u_{xx}^i = 0 & (t,x) \in \mathbb{R} \times [0, \ell_i], \\
u^i(t, \ell_i) = 0, \quad u^i(t,0) = v(t) & t \in \mathbb{R}, \\
u^i(0,x) = u_0^i(x), \quad u_t^i(0,x) = u_1^i(x) & x \in [0, \ell_i],
\end{cases}
\tag{4.10}
$$

for $i = 1, 2$. In this case the system is constituted by two uncoupled wave equations and the term *simultaneous* refers to the fact that the control v applied to both strings is the same. Chapter 5 of [78] is devoted to this sort of problems, considered for the first time by Russell in [107].

For every $T > 0$ system (4.10) is well posed for initial states $(u_0^i, u_1^i) \in L^2(0, \ell_i) \times H^{-1}(0, \ell_i)$, $i = 1, 2$ and control $v \in L^2(0, T)$: there exists a unique solution satisfying

$$
u^i \in C([0,T] : L^2(0, \ell_i)) \cap C^1([0,T] : H^{-1}(0, \ell_i)), \quad i = 1, 2.
$$

When $v \equiv 0$ system (4.10) becomes

$$
\begin{cases}
\phi_{tt}^i - \phi_{xx}^i = 0 & (t,x) \in \mathbb{R} \times [0, \ell_i], \\
\phi^i(t, \ell_i) = \phi^i(t,0) = 0 & t \in \mathbb{R}, \\
\phi^i(0,x) = \phi_0^i(x), \quad \phi_t^i(0,x) = \phi_1^i(x) & x \in [0, \ell_i],
\end{cases}
\tag{4.11}
$$

with $i = 1, 2$. It is constituted by two wave equations with homogeneous Dirichlet boundary conditions, which are *uncoupled*. Both equations are well posed for $(\phi_0^i, \phi_1^i) \in H_0^1(0, \ell_i) \times L^2(0, \ell_i)$ and the corresponding solutions are expressed by the formula

$$
\phi^i(t,x) = \sum_{n \in \mathbb{N}} (\phi_{0,n}^i \cos \sigma_n^i t + \frac{\phi_{1,n}^1}{\sigma_n^i} \sin \sigma_n^i t) \sin \sigma_n^i x, \quad i = 1, 2, \tag{4.12}
$$

where (σ_n^i) is the sequence of the square roots of the eigenvalues of the string $\mathbf{e}_i$:

$$\sigma_n^i = \frac{n\pi}{\ell_i}, \quad n \in \mathbb{N},$$

and $(\phi_{0,n}^i)$, $(\phi_{1,n}^i)$ are the sequences of the Fourier coefficients of ϕ_0^i, ϕ_1^i, respectively, in the orthogonal basis $(\sin \sigma_n^i x)$ of $L^2(0, \ell_i)$:

$$\phi_0^i(x) = \sum_{n \in \mathbb{N}} \phi_{0,n}^i \sin \sigma_n^i x, \qquad \phi_1^i(x) = \sum_{n \in \mathbb{N}} \phi_{1,n}^i \sin \sigma_n^i x, \qquad i = 1, 2.$$

The control problem in time T consists in *characterizing the initial states* (u_0^i, u_1^i), $i = 1, 2$, *of system* (4.10) *such that there exists* $v \in L^2(0, T)$ *with the property that the solutions* u^1, u^2 *of* (4.10) *satisfy*

$$u^i(T, x) = u_t^i(T, x) = 0, \qquad i = 1, 2,$$

for $x \in [0, \ell_i]$.

Let us observe that, though system (4.11) is constituted by two *uncoupled* equations, the fact that the same control is used generates coupling conditions, similar to those arising in the three string network. In fact, if we apply HUM, it turns out that the simultaneous control problem is equivalent to proving the following observability inequality for (4.11)

$$\int_0^T |\phi_x^1(t, 0) + \phi_x^2(t, 0)|^2 dt \geq \sum_{i=1,2} \sum_{n \in \mathbb{N}} (c_n^i)^2 ((\sigma_n^i \phi_{0,n}^i)^2 + (\phi_{1,n}^i)^2). \quad (4.13)$$

If there exist sequences of positive numbers (c_n^i), $i = 1, 2$ such that (4.13) is verified by all the solutions ϕ^1, ϕ^2 of (4.11) with initial states $(\phi_0^1, \phi_1^1) \in Z^1 \times Z^1$, $(\phi_0^2, \phi_1^2) \in Z^2 \times Z^2$, respectively, then, all the initial states (u_0^i, u_1^i), $i = 1, 2$, satisfying

$$\sum_{n \in \mathbb{N}} \frac{1}{(c_n^i)^2} (u_{0,n}^i)^2 + \sum_{n \in \mathbb{N}} \frac{1}{(\sigma_n^i c_n^i)^2} (u_{1,n}^i)^2 < \infty$$

are controllable in time T.

Note that in inequality (4.13) the observed quantity is a combination of the derivatives ϕ_x^i at $x = 0$.

Our main observability result for the solutions of (4.11) is as follows:

Theorem 4.3. *Let* $T^* = 2(\ell_1 + \ell_2)$. *The following inequalities take place*

$$\int_0^{T^*} |\phi_x^1(t, 0) + \phi_x^2(t, 0)|^2 dt \geq \ell_1 \sum_{n \in \mathbb{N}} \left(\sin \sigma_n^1 \ell_2\right)^2 \left((\sigma_n^1 \phi_{0,n}^1)^2 + (\phi_{1,n}^1)^2\right),$$

$$\int_0^{T^*} |\phi_x^1(t, 0) + \phi_x^2(t, 0)|^2 dt \geq \ell_2 \sum_{n \in \mathbb{N}} \left(\sin \sigma_n^2 \ell_1\right)^2 \left((\sigma_n^2 \phi_{0,n}^2)^2 + (\phi_{1,n}^2)^2\right),$$

for any solution of (4.11) *with initial states* $(\phi_0^i, \phi_1^i) \in Z^i \times Z^i$, $i = 1, 2$.

Proof. We prove the second inequality; the first one can be proved in a similar way.

Let us observe that, due to the $2\ell_1$-periodicity in time of the component ϕ^1 of the solution of (4.11), it follows that $\ell_1^- \phi_x^1(t,0) = 0$, where ℓ_1^- is the operator defined by (3.7) corresponding to ℓ_1. Then, if we apply Proposition 3.3 we obtain

$$\int_0^{T^*} |\phi_x^1(t,0) + \phi_x^2(t,0)|^2 dt \geq \int_{\ell_1}^{T^*-\ell_1} |\ell_1^- \phi_x^1(t,0) + \ell_1^- \phi_x^2(t,0)|^2 dt$$

$$= \int_{\ell_1}^{T^*-\ell_1} |\ell_1^- \phi_x^2(t,0)|^2 dt. \tag{4.14}$$

On the other hand, $\psi = \ell_1^- \phi^2$ is a solution of the equation

$$\psi_{tt} - \psi_{xx} = 0$$

in $\mathbb{R} \times [0, \ell_2]$ and thus, from Proposition 3.1, it follows

$$\int_{\ell_2}^{T^*-\ell_2} |\psi_x(t,0)|^2 dt \geq 4\mathbf{E}_\psi. \tag{4.15}$$

Taking into account that $\psi_x(t,0) = \ell_1^- \phi_x^2(t,0)$, from (4.14) and (4.15) we obtain

$$\int_0^{T^*} |\phi_x^1(t,0) + \phi_x^2(t,0)|^2 dt \geq 4\mathbf{E}_{\ell_1^- \phi^2}. \tag{4.16}$$

It just remains to compute the energy $\mathbf{E}_{\ell_1^- \phi^2}$. From (4.12) we obtain that

$$\ell_1^- \phi^2(t,x) = \sum_{n \in \mathbb{N}} (\phi_{0,n}^2 \ell_1^- \cos \sigma_n^2 t + \frac{\phi_{1,n}^2}{\sigma_n^2} \ell_1^- \sin \sigma_n^2 t) \sin \sigma_n^2 x. \tag{4.17}$$

In view of the relations

$$\ell_1^- \cos \sigma_n^2 t = \frac{1}{2}\left(\cos \sigma_n^2 (t+\ell_1) - \cos \sigma_n^2 (t-\ell_1)\right) = -\sin \sigma_n^2 \ell_1 \sin \sigma_n^2 t,$$

$$\ell_1^- \sin \sigma_n^2 t = \frac{1}{2}\left(\sin \sigma_n^2 (t+\ell_1) - \sin \sigma_n^2 (t-\ell_1)\right) = \sin \sigma_n^2 \ell_1 \cos \sigma_n^2 t,$$

equality (4.17) becomes

$$\ell_1^- \phi^2(t,x) = \sum_{n \in \mathbb{N}} \sin \sigma_n^2 \ell_1 \left(\frac{\phi_{1,n}^2}{\sigma_n^2} \cos \sigma_n^2 t - \phi_{0,n}^2 \sin \sigma_n^2 t\right) \sin \sigma_n^2 x.$$

If we apply formula (2.25) for the energy it follows

$$\mathbf{E}_{\ell_1^- \phi^2} = \frac{\ell_2}{4} \sum_{n \in \mathbb{N}} \left(\sin \sigma_n^2 \ell_1\right)^2 \left((\sigma_n^2 \phi_{0,n}^1)^2 + (\phi_{1,n}^2)^2\right).$$

It suffices to replace the latter expression in (4.16) to obtain the observability inequality of the theorem.

Remark 4.4. It is important to remark that the results of Theorem 4.3 are not enhanced if we take a larger observation time. Indeed, due to the ℓ_1-periodicity of ϕ^1 and the ℓ_2-periodicity of ϕ^2 in time we have

$$\ell_1^- \ell_2^- \left(\phi_x^1(t,0) + \phi_x^2(t,0)\right) = 0$$

for every $t \in \mathbb{R}$. This implies that, for every $T \geq T^*$ there exists a constant $C_T > 0$ such that

$$\int_0^T \left|\phi_x^1(t,0) + \phi_x^2(t,0)\right|^2 dt \leq C_T \int_0^{T^*} \left|\phi_x^1(t,0) + \phi_x^2(t,0)\right|^2 dt.$$

Consequently, if

$$\int_0^T \left|\phi_x^1(t,0) + \phi_x^2(t,0)\right|^2 dt$$

defines a norm in the space of initial states for system (4.11) for some $T \geq T^*$, so does

$$\int_0^{T^*} \left|\phi_x^1(t,0) + \phi_x^2(t,0)\right|^2 dt$$

and both norms are equivalent.

4.2.1 Identification of Controllable Subspaces

The aim of this subsection is to identify subspaces of controllable initial data of system (4.10) in time $T \geq 2(\ell_1 + \ell_2)$ with the aid of Theorem 4.3.

An easily identifiable subspace is that of the finite linear combinations of the eigenfunctions. The following holds

Proposition 4.5. *System (4.10) is spectrally controllable in some time $T \geq 2(\ell_1 + \ell_2)$ if, and only if, the quotient ℓ_1/ℓ_2 is an irrational number.*

Proof. If ℓ_1/ℓ_2 is irrational then the coefficients $\sin \sigma_n^1 \ell_2$, $\sin \sigma_n^2 \ell_1$, $n \in \mathbb{N}$, appearing in the inequalities in Proposition 4.3 are all different from zero. Indeed, if $\sin \sigma_n^1 \ell_2 = 0$ for some n, then there would exist $k \in \mathbb{N}$ such that

$$\frac{n\pi}{\ell_1} \ell_2 = k\pi,$$

that is, $\ell_1/\ell_2 = n/k \in \mathbb{Q}$. Then, the initial states (u_0^i, u_1^i), $i = 1, 2$, satisfying

$$\sum_{n \in \mathbb{N}} \frac{1}{(\sin \sigma_n^1 \ell_2)^2} (u_{0,n}^1)^2 + \sum_{n \in \mathbb{N}} \frac{1}{(\sigma_n^1 \sin \sigma_n^1 \ell_2)^2} (u_{1,n}^1)^2 < \infty, \qquad (4.18)$$

$$\sum_{n \in \mathbb{N}} \frac{1}{(\sin \sigma_n^2 \ell_1)^2} (u_{0,n}^2)^2 + \sum_{n \in \mathbb{N}} \frac{1}{(\sigma_n^2 \sin \sigma_n^2 \ell_1)^2} (u_{1,n}^2)^2 < \infty, \qquad (4.19)$$

are controllable in time $T \geq 2(\ell_1 + \ell_2)$. In particular, the initial states $(u_0^1, u_1^1) \in Z^1 \times Z^1$, $(u_0^2, u_1^2) \in Z^2 \times Z^2$ are controllable.

Let us now see that the condition $\ell_1/\ell_2 \notin \mathbb{Q}$ is also necessary for approximate controllability and consequently for spectral controllability. If $\ell_1/\ell_2 = n/k$ with $n, k \in \mathbb{N}$ then, for every $p \in \mathbb{N}$ the functions

$$\phi^1(t, x) = \sin \frac{pn\pi t}{\ell_1} \sin \frac{pn\pi x}{\ell_1}, \qquad \phi^2(t, x) = -\sin \frac{pk\pi t}{\ell_2} \sin \frac{pk\pi x}{\ell_2},$$

are solutions of (4.11) and satisfy

$$\phi_x^1(t, 0) + \phi_x^2(t, 0) \equiv 0.$$

Consequently, system (4.10) is not approximately controllable and, in particular, is not spectrally controllable.

To further pursue in the identification of controllable initial states of system (4.10) with the aid of Theorem 4.3 we need some definitions from Number Theory. For $\eta \in \mathbb{R}$ we denote by $|||\eta|||$ the distance from η to the set $\mathbb{Z}$:

$$|||\eta||| = |\min\{x \in \mathbb{R} : \eta - x \in \mathbb{Z}\}|.$$

Proposition 4.6. *If ℓ_1/ℓ_2 is irrational, then all the initial states (u_0^1, u_1^1), (u_0^2, u_1^2) satisfying*

$$\sum_{n \in \mathbb{N}} \frac{1}{|||n\frac{\ell_2}{\ell_1}|||^2} (u_{0,n}^1)^2 + \sum_{n \in \mathbb{N}} \frac{1}{n^2 |||n\frac{\ell_2}{\ell_1}|||^2} (u_{1,n}^1)^2 < \infty, \qquad (4.20)$$

$$\sum_{n \in \mathbb{N}} \frac{1}{|||n\frac{\ell_1}{\ell_2}|||^2} (u_{0,n}^2)^2 + \sum_{n \in \mathbb{N}} \frac{1}{n^2 |||n\frac{\ell_1}{\ell_2}|||^2} (u_{1,n}^2)^2 < \infty, \qquad (4.21)$$

are controllable in time $T \geq 2(\ell_1 + \ell_2)$.

Proof. Let us observe that for each $x \in \mathbb{R}$

$$2|||\frac{x}{\pi}||| \leq |\sin x| \leq \pi|||\frac{x}{\pi}||| \qquad (4.22)$$

(the proof of this fact may be found in Proposition A.1 in Appendix A).
Then,

$$2|||n\frac{\ell_2}{\ell_1}||| \leq |\sin \sigma_n^1 \ell_2| \leq \pi|||n\frac{\ell_2}{\ell_1}|||, \qquad 2|||n\frac{\ell_1}{\ell_2}||| \leq |\sin \sigma_n^2 \ell_1| \leq \pi|||n\frac{\ell_1}{\ell_2}|||.$$

Thus, relations (4.20)-(4.21) are equivalent to (4.18)-(4.19).

Therefore, in order to characterize subspaces of controllable initial states for (4.10) it suffices to estimate the norms of the sequences $|||n\frac{\ell_2}{\ell_1}|||$, $|||n\frac{\ell_1}{\ell_2}|||$, $n \in \mathbb{N}$.

A natural way of getting additional information is the following: let $\rho :$ $\mathbb{R} \to \mathbb{R}_+$ be an increasing function and define

$$\Psi_\rho = \left\{ x \in \mathbb{R}_+ : \liminf_{n \to \infty} |||nx||| \rho(n) > 0 \right\}.$$

Then, if $\ell_1/\ell_2, \ell_2/\ell_1 \in \Psi_\rho$ the inequalities

$$\sum_{n \in \mathbb{N}} \rho^2(n)(u_{0,n}^1)^2 + \sum_{n \in \mathbb{N}} \frac{\rho^2(n)}{n^2}(u_{1,n}^1)^2 < \infty, \qquad (4.23)$$

$$\sum_{n \in \mathbb{N}} \rho^2(n)(u_{0,n}^2)^2 + \sum_{n \in \mathbb{N}} \frac{\rho^2(n)}{n^2}(u_{1,n}^2)^2 < \infty, \qquad (4.24)$$

guarantee the controllability of the initial state (u_0^1, u_1^1), (u_0^2, u_1^2).

In what follows we restrict ourselves to the case $\rho(x) = x^\alpha$ with $\alpha > 0$. This choice is motivated by two reasons. The first one is that it leads to the identification of subspaces of controllable initial states of the form

$$(u_0^i, u_1^i) \in \hat{H}^\alpha(0, \ell_i) \times \hat{H}^{\alpha-1}(0, \ell_i),$$

where

$$\hat{H}^s(0, \ell_i) = \left\{ u(x) = \sum_{n \in \mathbb{N}} u_n \sin \sigma_n^i x : \sum_{n \in \mathbb{N}} n^{2s}(u_n)^2 < \infty \right\}.$$

Let us note that $\hat{H}^s(0, \ell_i)$ is the domain of the $s/2$-power of the laplacian and it is a closed subspace of the Sobolev space $H^s(0, \ell_i)$ with certain additional boundary conditions. In particular, $\hat{H}^1(0, \ell_i) = H_0^1(0, \ell_i)$ and $\hat{H}^0(0, \ell_i) = L^2(0, \ell_i)$.

The second reason for this choice of the function ρ is that the problem of describing the sets

$$\Psi_\alpha := \Psi_{x^\alpha} = \left\{ x \in \mathbb{R}_+ : \liminf_{n \to \infty} |||nx||| n^\alpha > 0 \right\},$$

is a classical and difficult one in Number Theory. In [69] and [26] the reader may find information on this topic. We refer also to Appendix A where we have gathered the most relevant facts, which will be used in what follows.

We now summarize the main consequences of our analysis distinguishing positive and negative results:

Positive results

The following results are known

1) For every $\alpha > 0$ the sets Ψ_α have the following property: *if $\xi \in \Psi_\alpha$ then $1/\xi \in \Psi_\alpha$.*

2) Ψ_1 coincides with the set of irrational numbers $\eta \in \mathbb{R}$ having a continuous fraction expansion $[a_0, a_1, ..., a_n, ...]$ (see, e.g., [69], p. 6) with bounded (a_n). The set Ψ_1 is not countable but has zero Lebesgue measure.

3) For every $\varepsilon > 0$ the complement of the set $\Psi_{1+\varepsilon}$ is of measure zero (see Proposition A.5 in Appendix A). This set is usually denoted in the literature as $\mathbf{B}_\varepsilon \subset \mathbb{R}$. As a consequence of Roth's theorem (Theorem A.4), the set $\mathbf{B}_\varepsilon$ contains all the algebraic irrational numbers, that is, all the roots of polynomials of degree greater than one with integer coefficients.

As a consequence we obtain

Corollary 4.7. *a) If $\ell_1/\ell_2 \in \mathbf{B}_\varepsilon$ then the subspace of initial states*

$$(u_0^i, u_1^i) \in \hat{H}^{1+\varepsilon}(0, \ell_i) \times \hat{H}^\varepsilon(0, \ell_i),$$

is controllable in any time $T \geq 2(\ell_1 + \ell_2)$. In particular, if ℓ_1/ℓ_2 is an algebraic irrational number, this subspace is controllable for any $\varepsilon > 0$.

b) If ℓ_1/ℓ_2 admits a bounded expansion in continuous fraction then, the subspace of initial states $(u_0^i, u_1^i) \in H_0^1(0, \ell_i) \times L^2(0, \ell_i)$, is controllable in any time $T \geq 2(\ell_1 + \ell_2)$.

Remark 4.8. Note that, in both cases, the space of controllable data is at most $H_0^1(0, \ell_i) \times L^2(0, \ell_i)$. It is important to remark that this space is smaller by one derivative than the controllable space for each individual string $L^2(0, \ell_i) \times H^{-1}(0, \ell_i)$. Therefore we see that, even though simultaneous controllability is possible under suitable assumptions on the legths ℓ_1, ℓ_2, it necessarily holds in a weaker space, with a loss of one derivative.

Negative results

We now describe some results on the lack of controllability that may be obtained as a consequence of the characterization in Proposition 4.6.

Proposition 4.9. *If there exists a sequence $(n_k) \subset \mathbb{N}$ such that*

$$\left|\left|\left|n_k \frac{\ell_1}{\ell_2}\right|\right|\right| \rho(n_k) \to 0 \quad or \quad \left|\left|\left|n_k \frac{\ell_2}{\ell_1}\right|\right|\right| \rho(n_k) \to 0, \qquad k \to \infty,$$

then, there exist initial states (u_0^1, u_1^1), (u_0^2, u_1^2) satisfying (4.23)-(4.24) which are not controllable in any finite time T.

Proof. Recall that the fact that all the initial states satisfying (4.23)-(4.24) are controllable in time T is equivalent to the inequalities:

$$\int_0^T |\phi_x^1(t, 0) + \phi_x^2(t, 0)|^2 dt \geq C_1 \sum_{n \in \mathbb{N}} \frac{1}{\rho^2(n)} \left((\frac{n\pi}{\ell_1} \phi_{0,n}^1)^2 + (\phi_{1,n}^1)^2 \right), \text{ (4.25)}$$

$$\int_0^T |\phi_x^1(t, 0) + \phi_x^2(t, 0)|^2 dt \geq C_2 \sum_{n \in \mathbb{N}} \frac{1}{\rho^2(n)} \left((\frac{n\pi}{\ell_2} \phi_{0,n}^2)^2 + (\phi_{1,n}^2)^2 \right), \text{ (4.26)}$$

for any solution of (4.11) with initial states $(\phi_0^i, \phi_1^i) \in Z^i \times Z^i$, $i = 1, 2$.

When

$$|||n_k \frac{\ell_1}{\ell_2}|||\rho(n_k) \to 0 \qquad (4.27)$$

inequality (4.26) is impossible. Indeed, from (4.27) it holds that, for every $k \in \mathbb{N}$, there exists $m_k \in \mathbb{N}$ such that

$$\left| n_k \frac{\ell_1}{\ell_2} - m_k \right| \rho(n_k) \to 0.$$

Then,

$$\left| \sigma_{n_k}^2 - \sigma_{m_k}^1 \right| \rho(n_k) = \left| \frac{\pi n_k}{\ell_2} - \frac{\pi m_k}{\ell_1} \right| \rho(n_k) \to 0. \qquad (4.28)$$

On the other hand, after replacing in (4.26) the solutions

$$\phi_k^1(t, x) = \cos \sigma_{m_k}^1 t \sin \sigma_{m_k}^1 x, \qquad \phi_k^2(t, x) = -\cos \sigma_{n_k}^2 t \sin \sigma_{n_k}^2 x,$$

it holds

$$\int_0^T |\sigma_{m_k}^1 \cos \sigma_{m_k}^1 t - \sigma_{n_k}^2 \cos \sigma_{n_k}^2 t|^2 dt \geq C_2 \rho^{-2}(n_k)(\sigma_{n_k}^2)^2$$

and then

$$\left| \sigma_{n_k}^2 - \sigma_{m_k}^1 \right|^2 \geq C\rho^{-2}(n_k). \qquad (4.29)$$

Here we have used the inequality

$$\int_0^T |x \cos xt - y \cos yt|^2 dt \leq 4|x - y|^2 x^2 T,$$

for $y \geq x \geq 1$, which is easily obtained using, for example, the mean value theorem.

Thus, from (4.29) we obtain

$$\left| \sigma_{n_k}^2 - \sigma_{m_k}^1 \right| \rho(n_k) \geq C,$$

what contradicts (4.28).

The first important consequence of Proposition 4.9 is based on the Dirichlet theorem: *for all $\alpha < 1$, $\xi \in \mathbb{R}$ and $\varepsilon > 0$ there exists an infinite number of values of n such that $|||n\xi|||n^\alpha < \varepsilon$* (see [26], Section I.5).

Corollary 4.10. *For all values ℓ_1, ℓ_2 of the lengths of the strings and every $\alpha < 1$ there exist initial states*

$$(u_0^i, u_1^i) \in \hat{H}^\alpha(0, \ell_i) \times \hat{H}^{\alpha-1}(0, \ell_i), \qquad i = 1, 2,$$

which are not controllable in any finite time T. In particular, there exist non-controllable initial states in $L^2(0, \ell_i) \times H^{-1}(0, \ell_i)$, and, consequently, system (4.10) is not exactly controllable in any finite time.

Remark 4.11. According to this corollary the positive result in Corollary 4.7 is sharp since, whatever ℓ_1/ℓ_2 is, the space of controllable data is at most $\hat{H}^1 \times \hat{L}^2$ and, therefore, there is a loss of, at least, one derivative.

However, there are irrational values of ℓ_1/ℓ_2 for which the space of controllable data can be as small as one wishes. The following result of negative character is based on a construction due to Liouville.

Let us consider the series

$$\xi = \sum_{k \in \mathbb{N}} 10^{-a_k}, \tag{4.30}$$

where (a_k) is an increasing sequence of natural numbers. Then, for each $p \in \mathbb{N}$,

$$\min_{m \in \mathbb{Z}} |\xi 10^{a_p} - m| = 10^{a_p} \sum_{k > p} 10^{-a_k} < 10^{a_p - a_{p+1}} \sum_{k \geq 0} 10^{-k} < 10^{1 + a_p - a_{p+1}}. \tag{4.31}$$

Let us assume that $\rho : \mathbb{R} \to \mathbb{R}_+$ is an increasing function. Fix $\varepsilon > 0$ and choose a sequence (a_k) that verifies, for every $k \in \mathbb{N}$,

$$10^{1 + a_k - a_{k+1}} < \frac{\varepsilon}{\rho(10^{a_k})}, \tag{4.32}$$

or equivalently,

$$a_{k+1} > 1 + a_k + \lg \frac{\rho(10^{a_k})}{\varepsilon}.$$

Then, in view of (4.31) and (4.32) the number ξ defined by (4.30) satisfies

$$\min_{m \in \mathbb{Z}} |\xi 10^{a_p} - m| < \frac{\varepsilon}{\rho(10^{a_p})}.$$

Thus, for the sequence of natural numbers $n_p = 10^{a_p}$, $p \in \mathbb{N}$, it holds

$$|||n_p \xi||| \rho(n_p) < \varepsilon.$$

Summarizing, it is possible to construct real numbers ξ, which are approximated by rational ones faster than any given order ρ.

From Proposition 4.9 it follows:

Corollary 4.12. *For any increasing function $\rho : \mathbb{R} \to \mathbb{R}_+$, there exist values of the lengths ℓ_1, ℓ_2 of the strings and initial data in the subspace defined by (4.23)-(4.24), which are not controllable in any finite time T. In other words, the subspace of controllable initial states may be arbitrarily small.*

Remark 4.13. Numbers of the form (4.30) are the so-called Liouville's numbers. The discovery of such numbers had a transcendental importance in the history of Mathematics: Liouville proved that, if ξ is an algebraic irrational number of order p (that is, ξ is a root of a polynomial of degree p with rational coefficients and there is no polynomial of smaller degree having that property) then, the inequality

$$|\xi n - m| < \frac{1}{n^{p-1}}$$

has no solutions $m, n \in \mathbb{N}$. Therefore, the numbers defined by (4.30) are not algebraic. This fact allowed to show for the first time the existence of non algebraic numbers.

Remark 4.14. It is important to remark the elementary character of the proof of Theorem 4.3 if we compare it with the previously mentioned approaches based on generalized Ingham inequalities. We do not need to analyze the spectral gap and to use Ingham type inequalities. Moreover, our technique provides the optimal observation time.

4.3 The Three String Network with One Controlled Node

The rest of this chapter is devoted to the study of the control problem for the three string network with a single exterior controlled node. It can be reduced to the previously studied problem on the simultaneous control of two strings (see also Section 4.8). We prefer however to develop another method to directly address the three string network since it also works for general tree-shaped networks, a topic that we shall develop in Chapter 5.

The motion of the network is described by the system

$$\begin{cases} u_{tt}^i - u_{xx}^i = 0 & \text{in } \mathbb{R} \times [0, \ell_i], \ i = 0, 1, 2, \\ u^0(t, 0) = u^1(t, 0) = u^2(t, 0) & t \in \mathbb{R} \\ u_x^0(t, 0) + u_x^1(t, 0) + u_x^2(t, 0) = 0 & t \in \mathbb{R} \\ u^0(t, \ell_0) = v(t), \quad u^i(t, \ell_i) = 0 & t \in \mathbb{R} \qquad i = 1, 2, \\ u^i(0, x) = u_0^i(x), \quad u_t^i(0, x) = u_1^i(x) \ x \in [0, \ell_i], \quad i = 0, 1, 2, \end{cases} \qquad (4.33)$$

which coincides with (4.1), except by the fact that now $v^1 = 0$. In other words, one single control is now entering on the system.

Let us observe that the homogeneous version of system (4.33), that is, when $v = 0$, coincides with (4.2).

From the general results of Chapter 2 we know that the observability inequality

$$\int_0^T |\phi_x^0(t, \ell_0)|^2 dt \geq \sum_{n \in \mathbb{N}} c_n^2 (\mu_n \phi_{0,n}^2 + \phi_{1,n}^2), \qquad (4.34)$$

for every solution $\bar{\phi}$ of (4.2) with initial state $(\bar{\phi}_0, \bar{\phi}_1) \in Z \times Z$, and a sequence (c_n) of positive weights, is equivalent to the fact that the space of initial states $(\bar{u}_0, \bar{u}_1)$ verifying

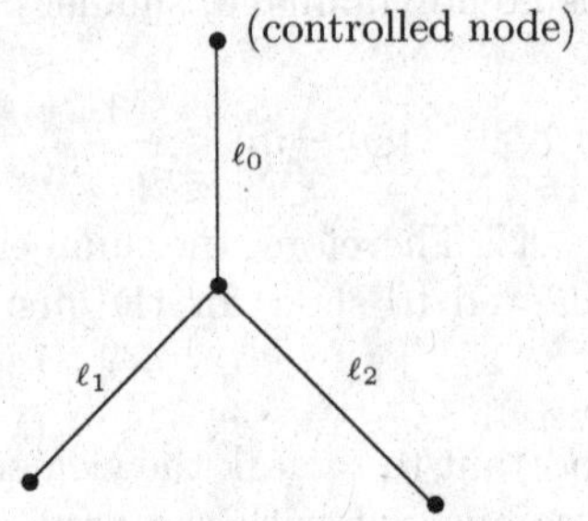

Fig. 4.3. The three string network with a single controlled node

$$\sum_{n \in \mathbb{N}} \frac{u_{0,n}^2}{c_n^2} < \infty, \qquad \sum_{n \in \mathbb{N}} \frac{u_{1,n}^2}{c_n^2 \mu_n} < \infty, \qquad (4.35)$$

is controllable in time T.

Then the exact controllability property of system (4.33) in $H \times V'$ is equivalent to the existence of a subsequence (c_n) satisfying (4.34) and having a positive lower bound:

$$c_n \geq c > 0, \qquad n \in \mathbb{N}. \qquad (4.36)$$

Unfortunately, that is impossible for system (4.2), whatever the values of the lengths ℓ_0, ℓ_1, ℓ_2 of the strings are. Indeed, as we shall see in Proposition 4.23, for all lengths ℓ_0, ℓ_1, ℓ_2, there exists a subsequence $(n_k) \subset \mathbb{N}$ such that

$$\lim_{k \to \infty} (\lambda_{n_k+1} - \lambda_{n_k}) = 0. \qquad (4.37)$$

Consequently, the infimum of the gap of consecutive eigenfrequencies entering in the Fourier development of solutions vanishes, and this for all values of ℓ_0, ℓ_1, and ℓ_2. We shall refer to this situation by saying simply that the spectral gap vanishes.

The spectral gap and the values of the weights (c_n) entering in (4.34) are closely related. Indeed, consider the solution

$$\bar{\phi}^k(t,x) = \frac{1}{\varkappa_{n_k+1}} \cos \lambda_{n_k+1} t \, \bar{\theta}_{n_k+1}(x) - \frac{1}{\varkappa_{n_k}} \cos \lambda_{n_k} t \, \bar{\theta}_{n_k}(x),$$

of (4.2), where

$$\varkappa_n = \theta_{n,x}^0(\ell_0).$$

Then

$$\phi_x^0(t,\ell_0) = \cos \lambda_{n_k+1} t - \cos \lambda_{n_k} t.$$

We have clearly,

$$\int_0^T |\cos\lambda_{n_k+1}t - \cos\lambda_{n_k}t|^2 dt \leq \frac{T^3}{3}|\lambda_{n_k+1} - \lambda_{n_k}|^2. \tag{4.38}$$

On the other hand, if (4.34) were true we would also have

$$\min(c_{n_k}, c_{n_k+1})\left(\frac{\lambda_{n_k+1}^2}{\varkappa_{n_k+1}^2} + \frac{\lambda_{n_k}^2}{\varkappa_{n_k}^2}\right) \leq \int_0^T |\cos\lambda_{n_k+1}t - \cos\lambda_{n_k}t|^2 dt$$

$$= \int_0^T |\phi_x^0(t,\ell_0)|^2 dt. \tag{4.39}$$

But $|\varkappa_n| \leq C\lambda_n$ (see formula (4.88) in Remark 4.31).

Then, as a consequence of (4.38) and (4.39) we have

$$\min(c_{n_k}, c_{n_k+1}) \leq C|\lambda_{n_k+1} - \lambda_{n_k}|^2,$$

for every k.

As a consequence, if the spectral gap condition is not satisfied and (4.37) holds, a sequence (c_n) such that the observaility property (4.34) holds is necessarily such that

$$\liminf_{n\to\infty} c_n = 0. \tag{4.40}$$

As we said above, whatever the values of the lengths ℓ_0, ℓ_1 and ℓ_2 are, the spectral gap of the three-string network vanishes. Thus, system (4.33) is not exactly controllable in the optimal space $H \times V'$ for any choice of the values of ℓ_0, ℓ_1, ℓ_2 and T. In this section we will prove inequalities of the form (4.34) for sequences of non-trivial weights (c_n) satisfying weakened positivity conditions. This will lead to controllability results in smaller spaces of initial data.

4.4 An Observability Inequality

In this section we prove the following property for the solutions of (4.2) which holds for all values of the lengths ℓ_i:

Theorem 4.15. *There exists a positive constant C such that every solution $\bar{\phi}$ of (4.2) with initial data $(\bar{\phi}_0, \bar{\phi}_1) \in V \times H$ satisfies the inequalities*

$$\int_0^{T^*} |\phi_x^0(t,\ell_0)|^2 dt \geq C\mathbf{E}_{\ell_j^-\bar{\phi}}, \quad j = 1,2, \tag{4.41}$$

where $T^* = 2(\ell_0 + \ell_1 + \ell_2)$.

In Theorem 4.15, $\ell_j^-\bar{\phi}$ stands for the function obtained by applying the operator ℓ^- defined in (3.7) for $\ell = \ell_j$ to the solution $\bar{\phi}$ of (4.2). Due to the linearity of ℓ_j^-, the function $\ell_j^- \phi$ is also a solution of (4.2). In particular, its energy $\mathbf{E}_{\ell_j^- \phi}(t)$ is conserved in time.

It is natural to try to proceed as in Section 4.1, that is, to apply Proposition 3.1 to estimate the energy of each string. This will allow us to show that, for every $\hat{t} \in \mathbb{R}$ and $i = 1, 2$,

$$\mathbf{E}_{\phi^i}(\hat{t}) \leq C \left(\int |\phi_x^i(t,0)|^2 dt + \int |\phi_t^i(t,0)|^2 dt \right),$$

$$\mathbf{E}_{\phi^0}(\hat{t}) \leq C \left(\int |\phi_x^0(t,\ell_0)|^2 dt + \int |\phi_t^0(t,\ell_0)|^2 dt \right) = \int |\phi_x^0(t,\ell_0)|^2 dt.$$

Note that we have not written the limits in the integrals explicitly. We will come back later to that issue in detail.

Thus, if we were able to prove that there exists $C > 0$ such that, for $i = 1, 2$,

$$\int |\phi_x^i(t,0)|^2 dt \leq C \int_0^T |\phi_x^0(t,\ell_0)|^2 dt, \tag{4.42}$$

$$\int |\phi_t^i(t,0)|^2 dt \leq C \int_0^T |\phi_x^0(t,\ell_0)|^2 dt, \tag{4.43}$$

we would obtain inequality (4.41).

Inequality (4.43) can be proved without difficulty for $i = 1, 2$: in view of the coupling conditions,

$$\phi_t^1(t,0) = \phi_t^2(t,0) = \phi_t^0(t,0) = \ell_0^- \phi_x^0(t,\ell_0), \tag{4.44}$$

and then

$$\int |\phi_t^i(t,0)|^2 dt = \int |\ell_0^- \phi_x^0(t,\ell_0)|^2 dt \leq \int |\phi_x^0(t,\ell_0)|^2 dt, \qquad i = 1, 2.$$

However, the inequalities (4.42) fail to hold[1]. In spite of what happens with $\phi_t^1(.,0)$, $\phi_t^2(.,0)$, the traces $\phi_x^1(.,0)$, $\phi_x^2(.,0)$ can not be expressed in a direct way as a function of $\phi_x^0(.,\ell_0)$; for them the coupling condition in this node allows simply deducing that

$$\phi_x^1(t,0) + \phi_x^2(t,0) = -\ell_0^+ \phi_x^0(t,\ell_0).$$

In other words, we get complete information on the linear combination $\phi_x^1(t,0) + \phi_x^2(t,0)$ but not on each individual string, i. e. on $\phi_x^1(t,0)$ and $\phi_x^2(t,0)$ separately.

Nevertheless, the boundary conditions $\phi_t^1(t,\ell_1) = \phi_t^2(t,\ell_2) = 0$ provide additional information that may be explicitly written as follows:

---

[1] If these inequalities were true, the whole energy space $H \times V'$ would be controllable and, as we have seen above, this is not true because of the lack of spectral gap.

$$0 = \phi_x^i(t, \ell_1) = \ell_i^+ \phi_t^i(t, 0) + \ell_i^- \phi_x^i(t, 0), \qquad i = 1, 2,$$

from where it holds

$$\ell_i^- \phi_x^i(t, 0) = -\ell_i^+ \phi_t^i(t, 0) = \ell_i^+ \ell_0^- \phi_x^0(t, \ell_0), \qquad i = 1, 2. \tag{4.45}$$

In this way, we arrive to the system of equations

$$\begin{cases} \phi_x^1(t, 0) + \phi_x^2(t, 0) = f(t), \\ \ell_1^- \phi_x^1(t, 0) = g_1(t), \qquad \ell_2^- \phi_x^2(t, 0) = g_2(t), \end{cases} \tag{4.46}$$

which is satisfied by the traces $\phi_x^1(., 0)$, $\phi_x^2(., 0)$, where

$$f(t) = -\ell_0^+ \phi_x^0(t, \ell_0), \qquad g_i(t) = \ell_i^+ \ell_0^- \phi_x^0(t, \ell_0), \qquad i = 1, 2.$$

Let us observe that f, g_1, g_2 are functions such that their norms in L^2 are bounded above in terms of the L^2-norm of $\phi_x^0(., \ell_0)$ with the help of Proposition 3.3.

As we shall see, the information (4.46) is sufficient to prove inequality (4.41). The idea of the proof is the following: if we apply, for example, the operator ℓ_1^- to the first equation in (4.46) we obtain the uncoupled conditions

$$\ell_1^- \phi_x^1(., 0) = g_1, \qquad \ell_1^- \phi_x^2(., 0) = \ell_1^- f - g_1. \tag{4.47}$$

Due to the linearity of system (4.2) and of the operators ℓ_1^- and ℓ_2^-, if $\bar{\phi}$ is a solution of (4.2) the functions $\ell_1^- \bar{\phi}$ and $\ell_2^- \bar{\phi}$ (the operators act in this case over the variable t) are also solutions of (4.2). Besides, the following identities take place

$$(\ell_j^- \phi^i)_x = \ell_j^- \phi_x^i, \quad (\ell_j^- \phi^i)_t = \ell_j^- \phi_t^i, \quad \text{for} \quad i = 0, 1, 2, \quad \text{and} \quad j = 1, 2.$$

Thus the solution $\bar{w} = \ell_1^- \bar{\phi}$ of (4.2) satisfies

$$(w^i)_x = \ell_1^- \phi_x^i, \quad (w^i)_t = \ell_1^- \phi_t^i, \quad \text{for} \quad i = 0, 1, 2,$$

and the relations (4.47) become

$$w_x^1(., 0) = g_1, \qquad w_x^2(., 0) = \ell_1^- f - g_1.$$

This implies that the L^2-norms of $w_x^1(., 0)$, $w_x^2(., 0)$, may be fully estimated in terms of the L^2-norm of $\phi_x^0(., \ell_0)$. Of course, the same happens with the traces $w_t^1(., 0)$, $w_t^2(., 0)$ and $w_x^0(., \ell_0)$ due to the continuity of ℓ_1^- (Proposition 3.3). With this it may be proved that

$$\int |\phi_x^0(t, \ell_0)|^2 dt \geq C \mathbf{E}_{\bar{w}}.$$

Following this simple argument we get:

Proposition 4.16. *There exists a positive constant C such that every solution $\bar{\phi}$ of (4.2) with initial data $(\bar{\phi}_0, \bar{\phi}_1) \in V \times H$ satisfies the inequalities*

$$\int_0^{T_j} |\phi_x^0(t, \ell_0)|^2 dt \geq C \mathbf{E}_{\ell_j^- \bar{\phi}}, \quad j = 1, 2,$$

where $T_j = 2(\ell_0 + \ell_j + \max\{\ell_1, \ell_2\})$.

Remark 4.17. This result is very close to Theorem 4.15 but in a larger observation time. If $\ell_1 < \ell_2$ the inequality of the theorem is immediately obtained for $j = 1$, since $T_1 = T^*$. But the time needed for $j = 2$ is larger in Proposition 4.16.

Proof. The proof of this Proposition is almost complete; we just need to follow carefully the integration intervals to obtain the indicated observation time. We will prove the assertion for $i = 1$; for $i = 2$ the proof is, obviously, similar.

Let us observe first that, since $w_t^0(t, \ell_0) = 0$, $w_x^0(t, \ell_0) = \ell_1^- \phi_x^0(t, \ell_0)$, then, in view of Proposition 3.1, for any $\hat{t} \in \mathbb{R}$, the energy $\mathbf{E}_{w^0}$ of the solution $\bar{w}$ on the string $\mathbf{e}_0$ satisfies

$$\mathbf{E}_{w^0}(\hat{t}) \leq C \int_{\hat{t}-\ell_0}^{\hat{t}+\ell_0} |w_x^0(t, \ell_0)|^2 dt = C \int_{\hat{t}-\ell_0}^{\hat{t}+\ell_0} |\ell_1^- \phi_x^0(t, \ell_0)|^2 dt,$$

and from Proposition 3.3 it follows that, for $\hat{t} \in [\ell_0 + \ell_1, T_1 - \ell_0 - \ell_1]$,

$$\mathbf{E}_{w^0}(\hat{t}) \leq C \int_{\hat{t}-\ell_0-\ell_1}^{\hat{t}+\ell_0+\ell_1} |\phi_x^0(t, \ell_0)|^2 dt \leq C \int_0^{T_1} |\phi_x^0(t, \ell_0)|^2 dt. \tag{4.48}$$

We claim that it is sufficient to prove the existence of $C > 0$ and $\hat{t} \in \mathbb{R}$ such that

$$\int_{\hat{t}-\ell_i}^{\hat{t}-\ell_i} |w_x^i(t, 0)|^2 dt \leq C \int_0^{T_1} |\phi_x^0(t, \ell_0)|^2 dt, \tag{4.49}$$

$$\int_{\hat{t}-\ell_i}^{\hat{t}-\ell_i} |w_t^i(t, 0)|^2 dt \leq C \int_0^{T_1} |\phi_x^0(t, \ell_0)|^2 dt, \tag{4.50}$$

$i = 1, 2$, for every solution of (4.2). Indeed, if this were true, then, based on Proposition 3.1 we would obtain

$$\mathbf{E}_{w^i}(\hat{t}) \leq C \int_0^{T_1} |\phi_x^0(t, \ell_0)|^2 dt, \quad i = 0, 1, 2$$

and then, in account of (4.48),

$$\mathbf{E}_{\bar{w}}(\hat{t}) = \mathbf{E}_{w^0}(\hat{t}) + \mathbf{E}_{w^1}(\hat{t}) + \mathbf{E}_{w^2}(\hat{t}) \leq C \int_0^{T_1} |\phi_x^0(t, \ell_0)|^2 dt.$$

So, we concentrate ourselves in proving inequalities (4.49)-(4.50). As it has been pointed out above in the formulas (4.44)-(4.45), we have the equalities

$$w_t^i(t,0) = \ell_i^- \phi_t^i(t,0) = \ell_i^- \ell_0^- \phi_x^0(t,\ell_0), \qquad i = 1,2, \qquad (4.51)$$

$$w_x^1(t,0) = \ell_1^- \phi_x^1(t,0) = -\ell_1^+ \phi_t^1(t,0), \ w_t^2(t,0) = w_t^1(t,0) = \ell_1^- \ell_0^- \phi_x^0(t,\ell_0). \tag{4.52}$$

Then, combining (4.51) with Proposition 3.3, we can ensure that, for any $\hat{t} \in \mathbb{R}$,

$$\int_{\hat{t}-\ell_1}^{\hat{t}+\ell_1} |w_t^1(t,0)|^2 dt = \int_{\hat{t}-\ell_1}^{\hat{t}+\ell_1} |\ell_1^- \ell_0^- \phi_x^0(t,0)|^2 dt \le C \int_{\hat{t}-\ell_0-2\ell_1}^{\hat{t}+\ell_0+2\ell_1} |\phi_x^0(t,\ell_0)|^2 dt.$$

In a similar way, the following inequalities hold

$$\int_{\hat{t}-\ell_1}^{\hat{t}+\ell_1} |w_x^1(t,0)|^2 dt \le C \int_{\hat{t}-\ell_0-2\ell_1}^{\hat{t}+\ell_0+2\ell_1} |\phi_x^0(t,\ell_0)|^2 dt,$$

$$\int_{\hat{t}-\ell_2}^{\hat{t}+\ell_2} |w_t^2(t,0)|^2 dt \le C \int_{\hat{t}-\ell_0-\ell_1-\ell_2}^{\hat{t}+\ell_0+\ell_1+\ell_2} |\phi_x^0(t,\ell_0)|^2 dt, \tag{4.53}$$

$$\int_{\hat{t}-\ell_2}^{\hat{t}+\ell_2} |w_x^2(t,0)|^2 dt \le C \int_{\hat{t}-\ell_0-\ell_1-\ell_2}^{\hat{t}+\ell_0+\ell_1+\ell_2} |\phi_x^0(t,\ell_0)|^2 dt.$$

Now it is easy to see that if we choose $\hat{t} = \ell_0 + \ell_1 + \max\{\ell_1,\ell_2\}$ in (4.53) we obtain[2] the inequalities (4.49)-(4.50). With this the proposition is proved.

Let us assume now that $\ell_1 \le \ell_2$. Then $T_1 = 2(\ell_0 + \ell_1 + \ell_2)$ and $T_2 = 2(\ell_0 + \ell_1 + \ell_2) \ge T_1$. But, in fact, the value of the observation time T_2 may be reduced.

The possibility of choosing an observation time smaller than T_2 (T_1 already coincides with T^*), which allows to obtain the assertion of the theorem from Proposition 4.16, is based on a property of *generalized periodicity* in time of the solutions of the homogeneous system (4.2) (see Proposition 4.18), which guarantees that all the L^2-information on the traces $\phi_x^0(t,\ell_0)$ of these solutions may be obtained in a time interval of length T^*. This makes observation in a larger time superfluous.

We define the operator

$$\mathcal{Q} := \ell_0^+ \ell_1^- \ell_2^- + \ell_0^- \ell_1^+ \ell_2^- + \ell_0^- \ell_1^- \ell_2^+. \tag{4.54}$$

Then,

---

[2] We have that $\hat{t} \in [\ell_0 + \ell_1, T_1 - \ell_0 - \ell_1]$, what is necessary for inequality (4.48). This value of $\hat{t}$ has been chosen so that the numbers $\hat{t} - \ell_0 - 2\ell_1$, $\hat{t} + \ell_0 + 2\ell_1$, $\hat{t} - \ell_0 - \ell_1 - \ell_2$, $\hat{t} + \ell_0 + \ell_1 + \ell_2$ belong to the interval $[0, T_1]$.

Proposition 4.18. *For every solution* $\bar{\phi}$ *of (4.2) with initial data* $(\bar{\phi}_0, \bar{\phi}_1) \in V \times H$ *it holds*

$$\mathfrak{Q}\phi_x^0(t, \ell_0) = 0.$$

Proof. From the relations $\phi_t^0(t,0) = -\ell_0^- \phi_x^0(t, \ell_0)$, $\phi_x^0(t,0) = \ell_0^+ \phi_x^0(t, \ell_0)$, we have that

$$\begin{aligned}
\mathfrak{Q}\phi_x^0(t, \ell_0) &= \ell_1^- \ell_2^- (\ell_0^+ \phi_x^0(t, \ell_0)) + (\ell_1^+ \ell_2^- + \ell_1^- \ell_2^+)\ell_0^- \phi_x^0(t, \ell_0) \\
&= \ell_1^- \ell_2^- \phi_x^0(t, 0) - (\ell_1^+ \ell_2^- + \ell_1^- \ell_2^+)\phi_t^0(t, 0).
\end{aligned}$$

Recalling that $\phi_t^0(t,0) = \phi_t^1(t,0) = \phi_t^2(t,0)$ we obtain

$$\mathfrak{Q}\phi_x^0(t, \ell_0) = \ell_1^- \ell_2^- \phi_x^0(t, 0) + \ell_2^-(-\ell_1^+ \phi_t^1(t,0)) + \ell_1^-(-\ell_2^+ \phi_t^2(t,0)). \quad (4.55)$$

Applying the equalities $\ell_1^+ \phi_t^1(t,0) + \ell_1^- \phi_x^1(t,0) = 0$, $\ell_2^+ \phi_t^2(t,0) + \ell_2^- \phi_x^2(t,0) = 0$ (obtained as in the proof of Proposition 4.16 from the boundary conditions) in (4.55) it holds

$$\begin{aligned}
\mathfrak{Q}\phi_x^0(t, \ell_0) &= \ell_1^- \ell_2^- \phi_x^0(t, 0) + \ell_2^- \ell_1^- \phi_t^1(t,0) + \ell_1^- \ell_2^- \phi_t^2(t,0) \\
&= \ell_1^- \ell_2^- (\phi_x^0(t,0) + \phi_t^1(t,0) + \phi_t^2(t,0)) = 0.
\end{aligned}$$

The usefulness of Proposition 4.18 in our context relies on the following property:

Lemma 4.19. *For every* $T > 0$ *there exists a constant* $C_T > 0$ *such that every continuous function* ψ, *which is a solution of* $Q\psi \equiv 0$, *satisfies the inequality*

$$\int_0^T |\psi(t)|^2 dt \le C_T \int_0^{T^*} |\psi(t)|^2 dt, \quad (4.56)$$

where, as before, $T^* = 2(\ell_0 + \ell_1 + \ell_2)$.

This fact, when applied to $\phi_x^0(t, \ell_0)$, yields the assertion of Theorem 4.15 from Proposition 4.16.

The proof of this property will be given in Chapter 5 for a larger class of operators $\mathfrak{Q}$. Now we restrict ourselves to the particular version corresponding to the operator $\mathfrak{Q}$ defined by (4.54), which allows to illustrate clearly the idea of the proof in the general case.

Let us denote $\ell_* = \min\{\ell_0, \ell_1, \ell_2\}$. We will prove that, for all $T > 0$ and ψ satisfying $Q\psi \equiv 0$,

$$\int_0^T |\psi(t)|^2 dt \le \int_0^{T^*} |\psi(t)|^2 dt + C \int_0^{T-2\ell_*} |\psi(t)|^2 dt. \quad (4.57)$$

From this inequality we can get (4.56), by iterating it as many times as necessary to obtain $T - 2n\ell_* \le T^*$. Indeed, in such case

$$\int_0^T |\psi(t)|^2 dt \le (1 + C + ... + C^{n-1}) \int_0^{T^*} |\psi(t)|^2 dt$$

$$+ C^n \int_0^{T - 2n\ell_*} |\psi(t)|^2 dt$$

$$\le (1 + ... + C^m) \int_0^{T^*} |\psi(t)|^2 dt, \tag{4.58}$$

which is the claimed inequality (4.56).

Let us observe that, according to the definition of $\mathcal{Q}$, the equality $\mathcal{Q}\psi \equiv 0$ may be written as

$$(\ell_0^+ \ell_1^- \ell_2^- + \ell_0^- \ell_1^+ \ell_2^- + \ell_0^- \ell_1^- \ell_2^+)\psi \equiv 0$$

and then, from the definition of the operators $\ell_i^\pm$, it follows that[3]

$$3\psi(t + \ell_0 + \ell_1 + \ell_2) - \psi(t + \ell_0 + \ell_1 - \ell_2) - \psi(t + \ell_0 - \ell_1 + \ell_2)$$
$$-\psi(t - \ell_0 - \ell_1 - \ell_2) - -\psi(t - \ell_0 + \ell_1 + \ell_2)$$
$$-\psi(t - \ell_0 + \ell_1 - \ell_2) - \psi(t - \ell_0 - \ell_1 + \ell_2) + 3\psi(t - \ell_0 - \ell_1 - \ell_2) = 0.$$

Replacing the variable t by $t - (\ell_0 + \ell_1 + \ell_2)$ the previous identity may be written as

$$\psi(t) = \sum_{\tau \in \Gamma} c_\tau \psi(t - \tau),$$

where

$$\Gamma := \{2\ell_0, 2\ell_1, 2\ell_2, 2(\ell_0 + \ell_1), 2(\ell_0 + \ell_2), 2(\ell_1 + \ell_2), 2(\ell_0 + \ell_1 + \ell_2)\}$$

with coefficients c_τ taking the values 1 or $-1/3$. From this, using the Cauchy-Schwarz inequality, we get

$$\int_{T^*}^T |\psi(t)|^2 dt \le C \sum_{\tau \in \Gamma} \int_{T^*}^T |\psi(t - \tau)|^2 dt = C \sum_{\tau \in \Gamma} \int_{T^* - \tau}^{T - \tau} |\psi(t)|^2 dt, \tag{4.59}$$

for some constant C independent of ψ (in fact, we may take $C = 55/9$).

But every $\tau \in \Gamma$ satisfies $2\ell^* \le \tau \le T^*$, so we have $T^* - \tau \ge 0$ and $T - \tau \le T - 2\ell_*$; and then the following inequality holds

$$\int_{T^* - \tau}^{T - \tau} |\psi(t)|^2 dt \le \int_0^{T - 2\ell_*} |\psi(t)|^2 dt.$$

Using the latter inequality in (4.59) we obtain

$$\int_{T^*}^T |\psi(t)|^2 dt \le C \sum_{\tau \in \Gamma} \int_0^{T - 2\ell_*} |\psi(t)|^2 dt = 7C \int_0^{T - 2\ell_*} |\psi(t)|^2 dt.$$

---

[3] It is a lengthly, but completely elementary computation.

Finally

$$\int_0^T |\psi(t)|^2 dt = \int_0^{T^*} |\psi(t)|^2 dt + \int_{T_*}^T |\psi(t)|^2 dt$$

$$\leq \int_0^{T^*} |\psi(t)|^2 dt + C \int_0^{T-2\ell_*} |\psi(t)|^2 dt, \qquad (4.60)$$

which is the inequality (4.57).

Thus, we have proved inequality (4.56) and with this, Theorem 4.15 is also proved.

4.5 Properties of the Sequence of Eigenvalues

The relation of the operator $\mathcal{Q}$ with system (4.2) is not purely technical. This operator is closely related to the boundary conditions (4.2), as Proposition 4.18 shows. Besides, the operator $\mathcal{Q}$ is linked in a direct way to the spectrum of $-\Delta_G$. Indeed, if we apply $\mathcal{Q}$ to the function $e^{i\lambda t}$, where λ is an arbitrary real number, since $\ell^+ e^{i\lambda t} = \cos \lambda \ell \; e^{i\lambda t}$ and $\ell^- e^{i\lambda t} = i \sin \lambda \ell \; e^{i\lambda t}$, we obtain

$$\mathcal{Q} e^{i\lambda t} = \left(\ell_0^+ \ell_1^- \ell_2^- + \ell_0^- \ell_1^+ \ell_2^- + \ell_0^- \ell_1^- \ell_2^+ \right) e^{i\lambda t}$$
$$= -\left(\cos \lambda \ell_0 \sin \lambda \ell_1 \sin \lambda \ell_2 + \sin \lambda \ell_0 \cos \lambda \ell_1 \sin \lambda \ell_2 \right.$$
$$\left. + \sin \lambda \ell_0 \sin \lambda \ell_1 \cos \lambda \ell_2 \right) e^{i\lambda t}.$$

Thus, we have

$$\mathcal{Q} e^{i\lambda t} = q(\lambda) e^{i\lambda t}$$

with

$$q(\lambda) := -(\cos \lambda \ell_0 \sin \lambda \ell_1 \sin \lambda \ell_2 + \sin \lambda \ell_0 \cos \lambda \ell_1 \sin \lambda \ell_2 + \sin \lambda \ell_0 \sin \lambda \ell_1 \cos \lambda \ell_2).$$
$$(4.61)$$

The following holds

Proposition 4.20. *Let* $\lambda \neq 0$. *Then* λ^2 *is an eigenvalue of* $-\Delta_G$ *if, and only if,* $q(\lambda) = 0$.

Proof. The necessity of this condition is immediate: if λ^2 is an eigenvalue with associated eigenfunction $\bar\theta$ then the function $\bar\phi(t,x) = e^{i\lambda t}\bar\theta(x)$ is a solution of (4.2). According to Proposition 4.18 we have

$$0 = \mathcal{Q}\phi_x^0(t, \ell_0) = \mathcal{Q} e^{i\lambda t}\theta_x^0(\ell_0) = q(\lambda)e^{i\lambda t}\theta_x^0(\ell_0).$$

From this inequality it holds $q(\lambda) = 0$ if $\theta_x^0(\ell_0) \neq 0$. On the other hand, if $\theta_x^0(\ell_0) = 0$ then the function θ^0, which is a solution of a second order ordinary differential equation satisfies $\theta^0(\ell_0) = \theta_x^0(\ell_0) = 0$, and this implies $\theta^0 \equiv 0$; in particular, $\theta^0(0) = 0$. From the continuity conditions at $x = 0$ we have that

$\theta^1(0) = \theta^2(0) = 0$. This means that λ^2 is also an eigenvalue of the strings $\mathbf{e}_1$ and $\mathbf{e}_2$ and therefore,

$$\sin \lambda \ell_1 = \sin \lambda \ell_2 = 0.$$

If we replace these equalities in (4.61), we obtain $q(\lambda) = 0$.

Now we will see that the condition $q(\lambda) = 0$ is also sufficient for λ^2 to be an eigenvalue. To do this we construct a non-zero eigenfunction $\bar{\theta}$ associated to λ^2. We look for the components of $\bar{\theta}$ in the form:

$$\theta^i(x) = a_i \sin \lambda(x - \ell_i), \quad i = 0, 1, 2. \tag{4.62}$$

This guarantees that the boundary conditions at the external nodes ($\theta^i(\ell_i) = 0$) are satisfied. The remaining boundary conditions at $x = 0$ lead to the linear system

$$a_0 \sin \lambda \ell_0 = a_1 \sin \lambda \ell_1 = a_2 \sin \lambda \ell_2, \tag{4.63}$$

$$a_0 \lambda \cos \lambda \ell_0 + a_1 \lambda \cos \lambda \ell_1 + a_2 \lambda \cos \lambda \ell_2 = 0, \tag{4.64}$$

whose determinant coincides with $\lambda q(\lambda)$. Thus, if $q(\lambda) = 0$ we can find numbers a_0, a_1, a_2, not all of them equal to zero, such that the function $\bar{\theta}$ defined by (4.62) is an eigenfunction.

Remark 4.21. The proof of the fact that the condition given in Proposition 4.20 is necessary may be done in a simpler way without using the operator $\mathcal{Q}$. Indeed, in view of the boundary conditions, an eigenfunction is necessarily of the form (4.62). If this eigenfunction does not vanish identically, the determinant of the liner system (4.63) is equal to zero. Thus, $q(\lambda) = 0$. However, we have used the operator $\mathcal{Q}$ because this is the natural technique we will use in Chapter 5 to address more general networks.

An important consequence of Proposition 4.20 is the following. Let us denote by (σ_n) the increasing sequence formed by the elements of the set

$$\Sigma = \frac{\pi}{\ell_0} \mathbb{N} \cup \frac{\pi}{\ell_1} \mathbb{N} \cup \frac{\pi}{\ell_2} \mathbb{N}. \tag{4.65}$$

The elements of Σ are the positive square roots of the eigenvalues of the decoupled strings with homogeneous Dirichlet boundary conditions. Let (λ_n) be the increasing sequence formed by the positive square roots of the eigenvalues of the network. Then, for every $n \in \mathbb{N}$,

$$\lambda_n < \sigma_n < \lambda_{n+1} < \sigma_{n+1}. \tag{4.66}$$

Indeed, in every interval (σ_n, σ_{n+1}) the function $q(\lambda)$ may be expressed as

$$q(\lambda) = h_1(\lambda) h_2(\lambda),$$

where

$$h_1(\lambda) = \sin \lambda \ell_0 \sin \lambda \ell_1 \sin \lambda \ell_2, \qquad h_2(\lambda) = \cot \lambda \ell_0 + \cot \lambda \ell_1 + \cot \lambda \ell_2.$$

Observe that under the hypothesis that all the ratios ℓ_i/ℓ_j ($i \neq j$) are irrational, the numbers λ_n are the positive zeros of $h_2(\lambda)$, while σ_n are the points where $h_2(\lambda) \to \pm\infty$. It suffices now to note that on every interval (σ_n, σ_{n+1}) the function $h_2(\lambda)$ is strictly increasing to conclude that, necessarily, the numbers σ_n and λ_n alternate, that is, (4.66) is verified.

Inequalities (4.66) allow to obtain information on the numbers λ_n from the properties of the sequence (σ_n). Indeed, observe that from (4.66) we obtain that, for every $n \in \mathbb{N}$,

$$\lambda_{n+4} - \lambda_n > \sigma_{n+3} - \sigma_n. \tag{4.67}$$

But, for every $n \in \mathbb{N}$, among the four numbers $\sigma_n, \sigma_{n+1}, \sigma_{n+2}, \sigma_{n+3}$ there are at least two corresponding to the same string. Consequently, for every $n \in \mathbb{N}$, there exists $i \in \{0, 1, 2\}$ such that

$$\sigma_{n+3} - \sigma_n > \frac{\pi}{\ell_i}.$$

Therefore, for every $n \in \mathbb{N}$,

$$\sigma_{n+3} - \sigma_n > \pi \min \left(\frac{1}{\ell_0}, \frac{1}{\ell_1}, \frac{1}{\ell_2} \right).$$

In other words, the following generalized separation property holds:

Proposition 4.22. *For every $n \in \mathbb{N}$ it holds*

$$\lambda_{n+4} - \lambda_n > \pi \min \left(\frac{1}{\ell_0}, \frac{1}{\ell_1}, \frac{1}{\ell_2} \right). \tag{4.68}$$

This generalized separation property (4.68) allows to apply the technique derived from Theorem 3.29 developed in [11] and [17] in the proof of observability inequalities of the type (4.34).

On the other hand, if n_λ and n_σ denote respectively, the counting functions[4] of the sequences (λ_n) and (σ_n) then

$$n_\sigma(r) \leq n_\lambda(r) \leq n_\sigma(r) + 1.$$

The function $n_\sigma(r)$ coincides with the sum of the counting functions of the sequences $(n\pi/\ell_0)$, $(n\pi/\ell_1)$ and $(n\pi/\ell_2)$. It holds

$$n_\sigma(r) = \left[\frac{r\ell_0}{\pi}\right] + \left[\frac{r\ell_1}{\pi}\right] + \left[\frac{r\ell_2}{\pi}\right],$$

where $[\eta]$ denotes the integer part of the number η. Therefore, we obtain

---

[4] The counting function $n(r)$ of the sequence of positive numbers (λ_n) is the number of elements of the sequence contained on the interval $(0, r]$.

$$r\frac{\ell_0 + \ell_1 + \ell_2}{\pi} - 3 \le n_\lambda(r) \le 1 + r\frac{\ell_0 + \ell_1 + \ell_2}{\pi}. \tag{4.69}$$

Then, the sequence (λ_n) has density

$$D\left(\lambda_n\right) = \lim_{r\to\infty} \frac{n_\lambda(r)}{r} = \frac{\ell_0 + \ell_1 + \ell_2}{\pi},$$

which coincides with the density of the sequence (σ_n). It is essentially due to this reason that the time of observation $T^* = 2(\ell_0 + \ell_1 + \ell_2) = 2\pi D\left(\lambda_n\right)$ appearing in Theorem 4.15 is optimal. Indeed, the Beurling-Malliavin theorem ensures that for a sequence (λ_n) with density D the sequence $(e^{it\lambda_n})$ is complete in $L^2(0, T)$ if $T < 2\pi D$. This implies that a non-trivial inequality of the type (4.34) cannot be proved for $T < 2\pi D$.

Let us also observe that inequality (4.69) implies that

$$\lim_{n\to\infty} \frac{\lambda_n}{n} = \frac{\pi}{\ell_0 + \ell_1 + \ell_2}. \tag{4.70}$$

This shows that the eigenvalues of the network behave asymptotically as the eigenvalues of a single string of length $\ell_0 + \ell_1 + \ell_2$.

Another important consequence of the inequalities (4.66) is the following:

Proposition 4.23. *For any values ℓ_0, ℓ_1, ℓ_2 of the lengths of the strings there exists a subsequence $(n_k) \subset \mathbb{N}$ such that*

$$\lim_{k\to\infty} (\lambda_{n_k+1} - \lambda_{n_k}) = 0.$$

Proof. According to Dirichlet's theorem on the simultaneous approximation of real numbers by rational ones (see [26], Section I.5), for every $\varepsilon > 0$ there exist infinitely many values of $k \in \mathbb{N}$ for which there exist natural numbers p_k, q_k such that

$$\left| k\frac{\ell_1}{\ell_0} - p_k \right| < \varepsilon, \qquad \left| k\frac{\ell_2}{\ell_0} - q_k \right| < \varepsilon.$$

Then

$$\left| \frac{\pi k}{\ell_0} - \frac{\pi p_k}{\ell_1} \right| < \varepsilon', \qquad \left| \frac{\pi k}{\ell_0} - \frac{\pi q_k}{\ell_2} \right| < \varepsilon', \tag{4.71}$$

where

$$\varepsilon' = \max\left\{ \frac{\pi\varepsilon}{\ell_1}, \frac{\pi\varepsilon}{\ell_2} \right\}.$$

Let $n_k \in \mathbb{N}$ be such that

$$\sigma_{n_k} = \min\left\{ \frac{\pi k}{\ell_0}, \frac{\pi p_k}{\ell_1}, \frac{\pi q_k}{\ell_2} \right\}.$$

Then, from (4.71) we obtain the inequalities

$$\sigma_{n_k+2} - \sigma_{n_k} < \varepsilon'.$$

But, in view of (4.66), the latter inequality implies

$$\lambda_{n_k+1} - \lambda_{n_k} < \varepsilon',$$

for infinitely many values of $k \in \mathbb{N}$.

Taking into account that ε' may be chosen arbitrarily small, the assertion of the proposition is obtained.

Remark 4.24. In a similar way we can prove that for any values ℓ_0, ℓ_1, ℓ_2 of the lengths of the strings there exists a subsequence $(n_k) \subset \mathbb{N}$ such that

$$\lim_{k \to \infty} (\lambda_{n_k+3} - \lambda_{n_k}) = 0.$$

Thus, four is the minimal spectral step to ensure the generalized separation property of Proposition 4.22 to hold.

4.6 Fourier Representation of the Observability Inequality

Our aim in this section is to express the inequalities of Theorem 4.15 in terms of the Fourier coefficients of the initial data of the solution $\bar{\phi}$ of (4.2). To do this, we have to characterize $\mathbf{E}_{\ell_j^-\bar{\phi}}$, $j = 1, 2$, in terms of those coefficients.

If $\bar{\phi}_0, \bar{\phi}_1 \in Z$, that is, if the sequences $(\phi_{0,n})$ and $(\phi_{1,n})$ are finite, then from the formula (4.4) it follows

$$\ell_j^- \bar{\phi}(x) = \sum_{n \in \mathbb{N}} (\phi_{0,n} \ell_j^- \cos \lambda_n t + \frac{\phi_{1,n}}{\lambda_n} \ell_j^- \sin \lambda_n t) \bar{\theta}_n(x). \tag{4.72}$$

But

$$\ell_j^- \cos \lambda_n t = \frac{1}{2} (\cos \lambda_n (t + \ell_j) - \cos \lambda_n (t - \ell_j)) = -\sin \lambda_n \ell_j \sin \lambda_n t,$$

$$\ell_j^- \sin \lambda_n t = \frac{1}{2} (\sin \lambda_n (t + \ell_j) - \sin \lambda_n (t - \ell_j)) = \sin \lambda_n \ell_j \cos \lambda_n t.$$

Replacing these relations in (4.72) we obtain

$$\ell_j^- \bar{\phi}(x) = \sum_{n \in \mathbb{N}} \sin \lambda_n \ell_j (\frac{\phi_{1,n}}{\lambda_n} \cos \lambda_n t - \phi_{0,n} \sin \lambda_n t) \bar{\theta}_n(x). \tag{4.73}$$

Using the formula (4.5) for the energy of $\ell_j^- \bar{\phi}$ we arrive to

$$\mathbf{E}_{\ell_j^- \bar{\phi}} = \sum_{n \in \mathbb{N}} \sin^2 \lambda_n \ell_j (\mu_n \phi_{0,n}^2 + \phi_{1,n}^2) \tag{4.74}$$

and therefore, Theorem 4.15 allows us to ensure that there exists a constant $C > 0$ such that the inequalities

$$\int_0^{T^*} |\phi_x^0(t,\ell_0)|^2 dt \geq C \sum_{n \in \mathbb{N}} \sin^2 \lambda_n \ell_j (\mu_n \phi_{0,n}^2 + \phi_{1,n}^2), \quad j = 1, 2,$$

are verified for every $\bar{\phi}_0, \bar{\phi}_1 \in Z$. Since $Z \times Z$ is dense in $V \times H$, this inequality is still valid for all $\bar{\phi}_0 \in V$, $\bar{\phi}_1 \in H$.

If we denote

$$c_n := \max\{|\sin \lambda_n \ell_1|, |\sin \lambda_n \ell_2|\} \tag{4.75}$$

we have:

Theorem 4.25. *There exists a positive constant C such that every solution $\bar{\phi}$ of (4.2) with initial state $(\bar{\phi}_0, \bar{\phi}_1) \in V \times H$ satisfies*

$$\int_0^{T^*} |\phi_x^0(t,\ell_0)|^2 dt \geq C \sum_{n \in \mathbb{N}} c_n^2 (\mu_n \phi_{0,n}^2 + \phi_{1,n}^2). \tag{4.76}$$

4.7 Study of the Weights c_n

Theorem 4.25 provides a satisfactory result as it allows to ensure the controllability of the subspace of initial data defined by (4.35). However, that subspace depends on the coefficients c_n.

Let us observe first that when the ratio ℓ_1/ℓ_2 is a rational number there exist infinitely many linearly independent eigenfunctions which vanish identically on the controlled string. They are constructed as follows. If $\ell_1/\ell_2 = p/q$ with $p, q \in \mathbb{Z}$ then, for $k \in \mathbb{Z}$ we define the functions $\bar{\psi}_k = (\psi_k^0, \psi_k^1, \psi_k^2)$ by

$$\psi_k^0 \equiv 0, \quad \psi_k^1 = \sin \frac{kp\pi x}{\ell_1}, \quad \psi_k^2 = -\sin \frac{kq\pi x}{\ell_2}.$$

This fact implies that when ℓ_1/ℓ_2 is rational we cannot obtain an inequality like the one given in Theorem 4.25, with other non-vanishing coefficients c_n, not necessarily those defined by (4.75). Indeed, the solutions of (4.2) defined by

$$\bar{\phi}_k = \sin \frac{kp\pi t}{\ell_1} \bar{\psi}_k$$

satisfy (see figure below)

$$\phi_{k,x}^0(t,\ell_0) \equiv 0.$$

Thus, the condition $\ell_1/\ell_2 \notin \mathbb{Q}$ is necessary for an inequality like (4.76) to hold. In fact this condition is also sufficient:

Proposition 4.26. *If the ratio ℓ_1/ℓ_2 is an irrational number, then all the coefficients c_n, $n \in \mathbb{N}$, defined by (4.75), are different from zero.*

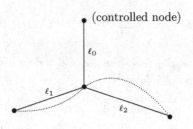

Fig. 4.4. A localized vibration when $\ell_2/\ell_1 \in \mathbb{Q}$.

Proof. It suffices to observe that $c_n = 0$ implies $|\sin \lambda_n \ell_1| = |\sin \lambda_n \ell_2| = 0$, and then

$$\lambda_n \ell_1 = p\pi, \qquad \lambda_n \ell_2 = q\pi.$$

An that is

$$\frac{\ell_1}{\ell_2} = \frac{p}{q} \in \mathbb{Q},$$

for some integers p and q. This contradicts the assumption $\ell_1/\ell_2 \notin \mathbb{Q}$.

Summarizing we have obtained the following characterization of the lengths of the strings for which the system is approximately or spectrally controllable in some finite time.

Corollary 4.27. *The following properties of the system of the three string network are equivalent:*

1) *The system is spectrally controllable in time $T \geq T^*$;*
2) *The system is approximately controllable in time $T \geq T^*$;*
3) *The ratio ℓ_1/ℓ_2 is an irrational number.*

The analysis of the previous section shows that for $T \geq T^*$ the space of observable and/or controllable states may be expressed in Fourier series by means of the weights c_n. In what follows we give sufficient conditions over the values of ℓ_0, ℓ_1, ℓ_2 allowing to ensure that for some $\alpha \in \mathbb{R}$ all the initial states $(\bar{u}_0, \bar{u}_1) \in \mathcal{W}^\alpha$ are controllable in time $T^* = 2(\ell_0 + \ell_1 + \ell_2)$. More precisely, if we define

$$\Phi_\alpha = \left\{ (\ell_0, \ell_1, \ell_2) \in \mathbb{R}_+^3 : \mathcal{W}^\alpha \subset \mathcal{W}_{T^*} \right\}$$

($\mathcal{W}_{T^*}$ is the subspace of controllable states in time T^*), our aim is to give explicit conditions guaranteeing that $(\ell_0, \ell_1, \ell_2) \in \Phi_\alpha$.

First, observe that if the weights c_n are such that for some $C > 0$

$$c_n \geq C\lambda_n^{-\alpha}, \tag{4.77}$$

$n \in \mathbb{N}$, then, as it has been pointed out in Section 4.3, $(\ell_0, \ell_1, \ell_2) \in \Phi_\alpha$.

Let us consider the function

$$\mathbf{a}^\alpha(\ell_1, \ell_2, \lambda) := (|\sin \lambda \ell_1| + |\sin \lambda \ell_2|) \lambda^\alpha.$$

It is also clear from (4.75) that, if for some values ℓ_1, ℓ_2 the function $\mathbf{a}^\alpha(\ell_1, \ell_2, \lambda)$ has a positive lower bound:

$$\mathbf{a}^\alpha(\ell_1, \ell_2, \lambda) \geq a > 0 \text{ for all } \lambda \in \mathbb{R}_+,$$

then, for every $\ell_0 \in \mathbb{R}_+$, inequality (4.77) holds. Thus, we will be concerned with the study of the function $\mathbf{a}$.

The following holds:

Proposition 4.28. *If there exists a constant $C > 0$ such that*

$$\left|\left|\left| n \frac{\ell_1}{\ell_2} \right|\right|\right| \geq C n^{-\alpha}$$

for every $n \in \mathbb{N}$, then

$$\mathbf{a}^\alpha(\ell_1, \ell_2, \lambda) \geq a > 0 \text{ for every } \lambda \geq 1. \tag{4.78}$$

Proof. Let us assume that the inequality (4.78) is false. Then there exists a sequence (λ_k) such that

$$\mathbf{a}^\alpha(\ell_1, \ell_2, \lambda_k) = |\sin \lambda_k \ell_1| \lambda_k^\alpha + |\sin \lambda_k \ell_2| \lambda_k^\alpha \to 0 \ (k \to \infty). \tag{4.79}$$

On the other hand, necessarily, $\lambda_k \to \infty$. Indeed, if (λ_k) has a finite limit point $\lambda_* \geq 1$ then, the continuity of $\mathbf{a}^\alpha$ yields

$$\mathbf{a}^\alpha(\ell_1, \ell_2, \lambda_*) = 0.$$

Consequently

$$\sin \lambda_* \ell_1 = \sin \lambda_* \ell_2 = 0,$$

and this may only happen if ℓ_1 / ℓ_2 is a rational number. But in such case there exist values of $n \in \mathbb{N}$ such that

$$\left|\left|\left| n \frac{\ell_1}{\ell_2} \right|\right|\right| = 0,$$

and this contradicts the hypothesis of the proposition. Thus we can assume that $\lambda_k \to \infty$.

From (4.79) it holds

$$|\sin \lambda_k \ell_1| \lambda_k^\alpha \to 0, \qquad |\sin \lambda_k \ell_2| \lambda_k^\alpha \to 0. \tag{4.80}$$

Let us denote for every $k \in \mathbb{N}$,

$$\varepsilon_k := \left|\left|\left| \frac{\ell_1 \lambda_k}{\pi} \right|\right|\right|, \qquad m_k := \frac{\ell_1 \lambda_k}{\pi} - \varepsilon_k \in \mathbb{N},$$

$$\delta_k := |||\frac{\ell_2\lambda_k}{\pi}|||, \qquad n_k := \frac{\ell_2\lambda_k}{\pi} - \delta_k \in \mathbb{N}.$$

Since $0 \leq \varepsilon_k, \delta_k \leq 1/2$,

$$\lim_{k\to\infty} \frac{m_k}{\lambda_k} = \frac{1}{\pi}, \qquad \lim_{k\to\infty} \frac{n_k}{\lambda_k} = \frac{1}{\pi}.$$

In particular, as $\lambda_k \to \infty$, the same happens with the sequences m_k and n_k. Besides,

$$\pi\varepsilon_k \leq 2|\sin\varepsilon_k\pi| = |\sin(m_k + \varepsilon_k)\pi| = \sin\lambda_k\ell_1,$$

and thus, from (4.80) we obtain $\varepsilon_k\lambda_k^\alpha \to 0$. Analogously, $\delta_k\lambda_k^\alpha \to 0$.

Then we have

$$\lim_{k\to\infty}\left(n_k\frac{\ell_1}{\ell_2} - m_k\right)n_k^\alpha = \frac{\pi}{\ell_2}\lim_{k\to\infty}\left(\varepsilon_k\frac{n_k}{\lambda_k} - \delta_k\frac{m_k}{\lambda_k}\right)n_k^\alpha$$
$$= \frac{1}{\ell_2}\lim_{k\to\infty}\left(\varepsilon_k n_k^\alpha - \delta_k n_k^\alpha\right) = 0. \qquad (4.81)$$

From this

$$|||n_k\frac{\ell_1}{\ell_2}|||n_k^\alpha \leq \left(n_k\frac{\ell_1}{\ell_2} - m_k\right)n_k^\alpha \to 0,$$

what contradicts the fact $|||n\ell_1/\ell_2||| \geq Cn^{-\alpha}$ for all $n \in \mathbb{N}$.

The condition provided by Proposition 4.28, which implies the inequality (4.77), is sufficient for $\mathcal{W}^\alpha \subset \mathcal{W}_T$ too. Now we will see that this condition is also necessary in the following sense

Proposition 4.29. *If there exists a sequence of natural numbers $n_k \to \infty$, for which*

$$|||n_k\frac{\ell_1}{\ell_2}|||n_k^\alpha \to 0, \ as \ k \to \infty,$$

then there exist values of $\ell_0 \in \mathbb{R}$ such that for every $T > 0$, the space $\mathcal{W}^\alpha$ is not contained in $\mathcal{W}_T$. That is, there exist initial states in $\mathcal{W}^\alpha$, which are not controllable.

Proof. It suffices to choose ℓ_0 such that

$$|||n_k\frac{\ell_0}{\ell_2}|||n_k^\alpha \to 0, \ as \ k \to \infty.$$

In fact, let m and $\tilde{m}$ be the closest natural numbers to $n_k\ell_1/\ell_2$ and $n_k\ell_0/\ell_2$, respectively. Let (σ_p) be the sequence defined by (4.65) and denote by p_k the index for which

$$\sigma_{p_k} = \min\left(n_k\frac{\pi}{\ell_2}, m\frac{\pi}{\ell_1}, \tilde{m}\frac{\pi}{\ell_0}\right).$$

Then we have

$$|\sigma_{p_k+2} - \sigma_{p_k}|\sigma_{p_k}^\alpha \to 0,$$

and this implies, in view of the inequalities (4.66),

$$|\lambda_{p_k+1} - \lambda_{p_k}|\lambda_{p_k}^\alpha \to 0. \tag{4.82}$$

On the other hand, the fact that all the initial states $(\bar{u}_0, \bar{u}_1) \in \mathcal{W}^\alpha$ are controllable in time T is equivalent to the following inequality

$$\int_0^T |\phi_x^0(t, \ell_0)|^2 dt \ge C \sum_{n \in \mathbb{N}} \lambda_n^{2\alpha}(\mu_n \phi_{0,n}^2 + \phi_{1,n}^2),$$

for all solution of (4.2) with $(\bar{\phi}_0, \bar{\phi}_1) \in Z \times Z$. However, proceeding as in Section 4.3, it can be easily proved that, due to (4.82), the latter inequality is impossible.

As in the case of the simultaneous control of two strings, Proposition 4.28 reduces the problem of identifying subspaces of controllable initial states to the following diophantine approximation problem: given $\alpha > 0$, to determine the values of ℓ_1/ℓ_2 for which there exists a constant $C > 0$ such that the inequality

$$|||n\ell_1/\ell_2||| \ge Cn^{-\alpha}, \tag{4.83}$$

is true for each $n \in \mathbb{N}$.

In view of the results described in Section 4.2.1 we obtain:

Corollary 4.30. *a) If $\ell_1/\ell_2 \in B_\varepsilon$ then, the space $\mathcal{W}^{1+\varepsilon}$ is controllable in any time $T \ge T^*$. In particular, if ℓ_1/ℓ_2 is an algebraic irrational number then, $\mathcal{W}^{1+\varepsilon}$ is controllable for any $\varepsilon > 0$.*

b) If ℓ_1/ℓ_2 admits a bounded expansion in continuous fraction then the subspace $\mathcal{W}^1$ is controllable in any time $T \ge T^*$.

c) There exist values of the lengths ℓ_0, ℓ_1, ℓ_2 such that no subspace of the form $\mathcal{W}^\alpha$ is controllable in finite time T.

Remark 4.31. As we will see later, the numbers $\varkappa_n = \theta_{n,x}^0(\ell_0)$, where $\bar{\theta}_n$ are the eigenfunctions of the elliptic problem associated to (4.2), are relevant for the control problem when we attempt to prove the observability inequalities in a direct way using Fourier series.

The eigenfunctions $\bar{\theta}_n$ may be explicitly expressed in terms of the eigenvalues λ_n,

$$\bar{\theta}_n = \begin{pmatrix} \theta_n^0 \\ \theta_n^1 \\ \theta_n^2 \end{pmatrix} = \gamma_n \begin{pmatrix} \dfrac{\sin \lambda_n (\ell_0 - x)}{\sin \lambda_n \ell_0} \\ \dfrac{\sin \lambda_n (\ell_1 - x)}{\sin \lambda_n \ell_1} \\ \dfrac{\sin \lambda_n (\ell_2 - x)}{\sin \lambda_n \ell_2} \end{pmatrix},$$

where

$$\gamma_n = \sqrt{2} \left\{ \frac{\ell_0}{\sin^2 \lambda_n \ell_0} + \frac{\ell_1}{\sin^2 \lambda_n \ell_1} + \frac{\ell_2}{\sin^2 \lambda_n \ell_2} \right\}^{-\frac{1}{2}}.$$

Then

$$\varkappa_n = -\lambda_n \sqrt{2} \left\{ \ell_0 + \ell_1 \frac{\sin^2 \lambda_n \ell_0}{\sin^2 \lambda_n \ell_1} + \ell_2 \frac{\sin^2 \lambda_n \ell_0}{\sin^2 \lambda_n \ell_2} \right\}^{-\frac{1}{2}}.$$

The following rough estimate is true

$$|\varkappa_n| \geq \lambda_n \sqrt{\frac{2}{\ell_0 + \ell_1 + \ell_2}} |\sin \lambda_n \ell_1 \sin \lambda_n \ell_2|. \tag{4.84}$$

If the lengths ℓ_0, ℓ_1, ℓ_2 are linearly independent over $\mathbb{Q}$ and the ratios ℓ_i/ℓ_j are algebraic numbers (see condition (S) in Appendix A for more details) then, according to Proposition A.11 from Appendix A, for each $\varepsilon > 0$ there exists a constant $C_\varepsilon > 0$ such that

$$|\sin \lambda_n \ell_1| \geq \frac{C_\varepsilon}{\lambda_n^{1+\varepsilon}}, \qquad |\sin \lambda_n \ell_2| \geq \frac{C_\varepsilon}{\lambda_n^{1+\varepsilon}}, \quad n \in \mathbb{N}, \tag{4.85}$$

and with this, from (4.84) it follows

$$|\varkappa_n| \geq \frac{C_\varepsilon}{\lambda_n^{1+\varepsilon}}. \tag{4.86}$$

However, with the aid of Theorem 4.25 it is possible to establish more precise estimates under weaker conditions on the lengths. Indeed, it suffices to apply the inequality of this theorem to the solutions $\sin \lambda_n t \, \bar\theta_n$ and $\cos \lambda_n t \, \bar\theta_n$ to get

$$|\varkappa_n|^2 \int_0^T |\cos \lambda_n t|^2 \, dt \geq C c_n^2 \lambda_n^2, \qquad |\varkappa_n|^2 \int_0^T |\sin \lambda_n t|^2 \, dt \geq C c_n^2 \lambda_n^2.$$

From this we obtain

$$|\varkappa_n| \geq C \lambda_n \max\{|\sin \lambda_n \ell_1|, |\sin \lambda_n \ell_2|\}.$$

This inequality is obviously sharper than (4.84). Consequently, if the ratio ℓ_1/ℓ_2 belongs to $\mathbf{B}_\varepsilon$ then,

$$|\varkappa_n| \geq \frac{C}{\lambda_n^\varepsilon}. \tag{4.87}$$

In spite of estimate (4.84), the latter does not impose any restriction over ℓ_0.

Proceeding in a similar way, from the inequality (2.29) we obtain

$$C \lambda_n \geq |\varkappa_n|, \tag{4.88}$$

independently of the values of the lengths.

Note that these two inequalities (4.87) and (4.88) reflect correctly the defect of at least one derivative on the boundary observation of the energy of individual eigenfunctions from the boundary measurement made on one single external node.

4.8 Relation Between the Simultaneous Control of Two Strings and the Control of the Three String Network from One Exterior Node

As the reader has noticed already, the conditions on the lengths of the strings that allow identifying subspaces of controllable initial states are the same for the simultaneous control of two strings and for the control of the three string network from one exterior node. Besides, when these conditions are satisfied, the corresponding subspaces of controllable initial states coincide, up to the boundary conditions, on the uncontrolled strings. There is indeed a very close connection between these two problems that explains this analogy:

Theorem 4.32. *If $\mathcal{V}$ is a subspace of controllable initial states in time T for the simultaneous control of two strings (4.10) then the subspace $(L^2(0,\ell_0) \times H^{-1}(0,\ell_0)) \times \mathcal{V}$ of initial states for the system of the three string network (4.33) is controllable in time $T + 2\ell_0$.*

For the proof of this fact we need some preliminary elements. We consider the spaces

$$\mathcal{W}_0 = \left\{ (\bar{\phi}_0, \bar{\phi}_1) \in V \times H : \quad \phi_0^0(0) = \phi_0^1(0) = \phi_0^2(0) = 0 \right\},$$
$$\mathcal{V}_0 = \left(H_0^1(0,\ell_1) \times L^2(0,\ell_1) \right) \times \left(H_0^1(0,\ell_2) \times L^2(0,\ell_2) \right).$$

For $(\bar{\phi}_0, \bar{\phi}_1) \in \mathcal{W}_0$ we denote by $\bar{\phi}$ the solution of the homogeneous system for the three string network (4.2) with initial state $(\bar{\phi}_0, \bar{\phi}_1)$. We also consider $\bar{\psi} = (0, \psi^1, \psi^2)$, where (ψ^1, ψ^2) is the solution of the homogeneous system (4.11) with initial states (ϕ_0^1, ϕ_1^1) and (ϕ_0^2, ϕ_1^2), respectively.

Let us choose $T \geq 2(\ell_0 + \ell_1 + \ell_2)$ and denote

$$||(\bar{\phi}_0, \bar{\phi}_1)||_E^2 := \int_0^T |\phi_x^0(t, \ell_0)|^2 dt,$$

$$||(\bar{\phi}_0, \bar{\phi}_1)||_S^2 := \int_{\ell_0}^{T-\ell_0} |\psi_x^1(t, 0) + \psi_x^2(t, 0)|^2 dt.$$

In view of the results of Proposition 4.5 and Corollary 4.27, the functions $||.||_E$ and $||.||_S$ define norms in $\mathcal{W}_0$ and $\mathcal{V}_0$ respectively, if, and only if, ℓ_1/ℓ_2 is an irrational number and $T \geq 2(\ell_0 + \ell_1 + \ell_2)$.

Proposition 4.33. *There exists a constant $C > 0$ such that for every $(\bar{\phi}_0, \bar{\phi}_1) \in \mathcal{W}_0$,*

$$C||(\bar{\phi}_0, \bar{\phi}_1)||_E^2 \geq ||(\phi_0^0, \phi_1^0)||_{H_0^1(0,\ell_0) \times L^2(0,\ell_0)}^2 + ||(\bar{\phi}_0, \bar{\phi}_1)||_S^2.$$

Proof. Let us observe that, if we apply D'Alembert formulas (3.5) to the component ϕ^0 we have, in account of the fact that $\phi_t^0(t, \ell_0) \equiv 0$,

$$\phi_x^0(t, 0) = \ell_0^- \phi_x^0(t, \ell_0), \qquad \phi_t^0(t, 0) = -\ell_0^+ \phi_x^0(t, \ell_0).$$

Then, from Proposition 3.3 we obtain the inequality

$$\int_0^T |\phi_x^0(t,\ell_0)|^2 dt \geq \max\{\int_{\ell_0}^{T-\ell_0} |\phi_x^0(t,0)|^2 dt, \int_{\ell_0}^{T-\ell_0} |\phi_t^0(t,0)|^2 dt\}. \quad (4.89)$$

On the other hand, if $\mathbf{E}_{\phi^0}$ is the energy of the component ϕ^0, from Proposition 3.1 it follows,

$$\mathbf{E}_{\phi^0}(0) \leq \int_{-\ell_0}^{\ell_0} |\phi_x^0(t,\ell_0)|^2 dt.$$

And then, from property (4.56) (see Proposition 4.18)

$$\|(\phi_0^0, \phi_1^0)\|_{H_0^1 \times L^2}^2 = 2\mathbf{E}_{\phi^0}(0) \leq C \int_0^T |\phi_x^0(t,\ell_0)|^2 dt. \quad (4.90)$$

Let us observe now that the solution $\bar{\phi}$ may be decomposed as

$$\bar{\phi} = \bar{\psi} + \bar{\omega}, \quad (4.91)$$

were $\bar{\omega} = (\omega^0, \omega^1, \omega^2)$ is the unique solution of the problem

$$\begin{cases} \omega_{tt}^i - \omega_{xx}^i = 0 & \text{in } \mathbb{R} \times [0, \ell_i], \quad i = 0, 1, 2, \\ \omega^i(t, \ell_i) = 0, \quad \omega^i(t, 0) = \phi^i(t, 0) & \text{in } \mathbb{R}, \quad\quad i = 0, 1, 2, \\ \omega^0(0, x) = \phi_0^0(x), \quad \omega_t^0(0, x) = \phi_1^0(x) \text{ in } [0, \ell_0], \\ \omega^i(0, x) = \omega_t^i(0, x) = 0, & \text{in } [0, \ell_i], \quad i = 1, 2. \end{cases} \quad (4.92)$$

Indeed, for every $i = 0, 1, 2$, the function $\bar{\eta} = \bar{\phi} - \bar{\psi} - \bar{\omega}$ satisfies

$$\begin{cases} \eta_{tt}^i - \eta_{xx}^i = 0 & \text{in } \mathbb{R} \times [0, \ell_i], \\ \eta^i(t, 0) = \eta^i(t, \ell_i) = 0 \text{ in } \mathbb{R}, \\ \eta^i(0, x) = \eta_t^i(0, x) = 0 \text{ in } [0, \ell_i]. \end{cases}$$

Thus, $\bar{\eta} \equiv \bar{0}$, that is, (4.91) is verified. In particular,

$$\phi_x^i(t, 0) = \psi_x^i(t, 0) + \omega_x^i(t, 0), \quad i = 0, 1, 2. \quad (4.93)$$

In view of the coupling conditions

$$\phi_x^0(t, 0) + \phi_x^1(t, 0) + \phi_x^2(t, 0) = 0,$$

from (4.93) it follows that

$$-\phi_x^0(t, 0) = \psi_x^1(t, 0) + \psi_x^2(t, 0) + \omega_x^1(t, 0) + \omega_x^2(t, 0). \quad (4.94)$$

Thus, we have

$$\int_{\ell_0}^{T-\ell_0} |\psi_x^1(t,0) + \psi_x^2(t,0)|^2 dt$$

$$\leq \int_{\ell_0}^{T-\ell_0} |\psi_x^1(t,0) + \psi_x^2(t,0) + \omega_x^1(t,0) + \omega_x^2(t,0)|^2 dt$$

$$+ \int_{\ell_0}^{T-\ell_0} |\omega_x^1(t,0) + \omega_x^2(t,0)|^2 dt$$

$$\leq \int_{\ell_0}^{T-\ell_0} |\phi_x^0(t,0)|^2 dt + \int_{\ell_0}^{T-\ell_0} |\omega_x^1(t,0) + \omega_x^2(t,0)|^2 dt.$$

On the other hand, if we apply Lemma 4.2 from [51] to the system (4.92), we obtain that there exists a constant $C > 0$ such that

$$\int_{\ell_0}^{T-\ell_0} |\omega_x^1(t,0) + \omega_x^2(t,0)|^2 dt \leq C \int_{\ell_0}^{T-\ell_0} |\omega_t^0(t,0)|^2 dt = C \int_{\ell_0}^{T-\ell_0} |\phi_t^0(t,0)|^2 dt.$$

So, we arrive to the inequality

$$\int_{\ell_0}^{T-\ell_0} |\psi_x^1(t,0) + \psi_x^2(t,0)|^2 dt \leq \int_{\ell_0}^{T-\ell_0} |\phi_x^0(t,0)|^2 dt + C \int_{\ell_0}^{T-\ell_0} |\phi_t^0(t,0)|^2 dt,$$

and, in view of (4.89),

$$\int_{\ell_0}^{T-\ell_0} |\psi_x^1(t,0) + \psi_x^2(t,0)|^2 dt \leq C \int_0^T |\phi_x^0(t,\ell_0)|^2 dt. \tag{4.95}$$

Finally, combining the inequalities (4.90) and (4.95) the assertion of the proposition is obtained.

Proposition 4.34. *Let $\bar{g} \in H$ be a continuous function such that $g^0(0) \neq 0$. Then, there exists a constant $C > 0$ such that for every $(\bar{\phi}_0, \bar{\phi}_1) \in W_0$ and every $\lambda \in \mathbb{R}$,*

$$C\|(\bar{\phi}_0 + \lambda\bar{g}, \bar{\phi}_1)\|_E \geq \|(\phi_0^0, \phi_1^0)\|_{H_0^1 \times L^2} + \|(\bar{\phi}_0, \bar{\phi}_1)\|_S.$$

Proof. Let us denote by $\bar{\varphi}_\lambda$ the solution of system (4.33) with initial state $(\bar{\phi}_0 + \lambda\bar{g}, \bar{\phi}_1)$. Let us observe that

$$|\lambda|^2 = |\phi_0^0(0) + \lambda g^0(0)|^2 \leq C\|\phi_0^0 + \lambda g^0\|_{H^1}^2 \leq C\mathbf{E}_{\varphi_\lambda^0}(0). \tag{4.96}$$

As in the proof of Proposition 4.33 we may show that

$$\mathbf{E}_{\varphi_\lambda^0}(0) \leq C\|(\bar{\phi}_0 + \lambda\bar{g}, \bar{\phi}_1)\|_E^2. \tag{4.97}$$

Then, from the relations (4.96), (4.97) we have

$$\|(\bar{\phi}_0, \bar{\phi}_1)\|_E \leq \|(\bar{\phi}_0 + \lambda, \bar{\phi}_1)\|_E + |\lambda|\|(\bar{g}, \bar{0})\|_E \leq C\|(\bar{\phi}_0 + \lambda\bar{g}, \bar{\phi}_1)\|_E$$

and the assertion holds from Proposition 4.33.

Proof (Proof of Theorem 4.32). Let us denote by $\mathcal{F}_S$ and $\mathcal{F}_E$ the completions of H and $\mathcal{V}_0$ with the norms $\|.\|_E$ and $\|.\|_S$, respectively. In account of the fact that

$$H = \mathbb{R}\bar{g} + \mathcal{W}_0,$$

Proposition 4.33 allows us to ensure that

$$\left(H_0^1(0, \ell_0) \times L^2(0, \ell_0)\right) \times \mathcal{F}_S \supset \mathcal{F}_E.$$

Then, the spaces of controllable initial states $\mathcal{C}_E = \mathcal{F}'_E$, $\mathcal{C}_S = \mathcal{F}'_S$ of systems (4.33) and (4.10) given by HUM satisfy the relation

$$\mathcal{C}_E \subset (L^2(0, \ell_0) \times H^{-1}(0, \ell_0)) \times \mathcal{C}_S.$$

In particular, if $\mathcal{V} \subset \mathcal{C}_S$ then

$$\mathcal{V} \subset (L^2(0, \ell_0) \times H^{-1}(0, \ell_0)) \times \mathcal{V},$$

and this is the assertion of the theorem.

Remark 4.35. The advantage of this approach that consists in deducing the controllability of the three string network from the simultaneous control of two strings is that:

a) It provides subspaces of controllable initial states of system (4.33) in which no restriction is imposed on the regularity of the components (u_0^0, u_1^0), other than being in $L^2(0, \ell_0) \times H^{-1}(0, \ell_0)$, in spite of what is needed in Corollary 4.30;

b) It is clear that no restriction on the length ℓ_0 is needed provided the control time is large enough ($T \geq 2(\ell_0 + \ell_1 + \ell_2)$). This is in agreement with common sense. Indeed, one expects that the initial data over the string whose external node is being controlled should be controlled correctly and that only further requirements should be needed on the data over the other two strings. Obviously, for that to be the case one has to take into account the extra time ($2\ell_0$) that controlling the third string $(0, \ell_0)$ adds.

4.9 Lack of Observability in Small Time

Due to the finite speed of propagation of waves along the strings of the network (equal to one in this case), it is natural to expect that, when the control time T is small the system is neither controllable nor observable. It turns out that this occurs whenever $T < T^* = 2(\ell_0 + \ell_1 + \ell_2)$. Consequently, the control time $T^* = 2(\ell_0 + \ell_1 + \ell_2)$ obtained in previous sections turns out to be sharp.

For an arbitrary network, the lack of spectral controllability for values of T smaller than twice its lengths may be proved on the basis of results from the Theory on Non Harmonic Fourier Series (more precisely, the Beurling-Malliavin theorem) and the asymptotic properties of the sequence of eigenvalues of the problem (see Chapter 6). However, for the three string network it is possible to give a completely elementary proof of this fact based on the explicit construction (shown in Figure 4.5) of a solution $\bar{\phi}$ of (4.2), whose trace $\phi_x^0(., \ell_0)$ in the observation point vanishes during a time $T < T^*$. This allows to ensure that the system (4.33) is not even approximately controllable $T < T^*$.

In Figure 4.5 we draw the projection of the support of the solution to the uncontrolled strings $(0, \ell_1)$ and $(0, \ell_2)$. The Figure shows a family of rays crossing each other at the connecting node $x = 0$ so that waves cancel when crossing at $x = 0$. This can be done in any time $\tau < 2(\ell_1 + \ell_2)$. One can then extend this solution by zero to the observed string $(0, \ell_0)$ guaranteeing that the observed trace at $x = \ell_0$ vanishes for a time interval of length $T < 2(\ell_0 + \ell_1 + \ell_2)$.

The construction is thus the same as that of the lack of simultaneous approximate controllability of two strings of lengths a and b from a common end-point in time less than $T < 2(a + b)$. This is precisely the situation we represent in Figure 4.5. When $T \geq 2(a + b)$ this construction can only be performed when a/b is rational (an then this may be done for all time $T > 0$). This is in agreement with our results on the simultaneous controllability of two strings in section 4.2 that guarantee approximate and spectral controllability as soon as a/b is irrational and $T \geq 2(a + b)$.

The following holds

Theorem 4.36. *Let $T < T^*$. Then, there exist non-zero initial states*

$$(\bar{\phi}_0, \bar{\phi}_1) \in \bigcap_{\alpha \in \mathbb{R}} \mathcal{W}^\alpha,$$

for which the solution $\bar{\phi}$ of (4.2) satisfies

$$\phi_x^0(t, \ell_0) = 0 \quad in \ [0, T]. \tag{4.98}$$

In the proof of this theorem we use some technical results. Let $T > 0$ and $0 < \sigma < T$. We define the operator $I_\sigma : L^2(0, T) \to L^2(0, T - \sigma)$ by the formula

$$(I_\sigma f)(t) := \int_t^{t+\sigma} f(\tau) d\tau.$$

For arbitrary values of $\sigma_1, \sigma_2 \in (0, T)$ the system of functional equations

$$\begin{cases} I_{\sigma_i} f_i = 0 & \text{a. e. in } (0, T - \sigma_i) \quad i = 1, 2, \\ f_1 + f_2 = 0 & \text{a. e. in } (0, T), \end{cases} \tag{4.99}$$

admits the trivial solution $f_1 = f_2 = 0$. We need to study for which values of T this is the only solution of (4.99). The answer is given by the following

Lemma 4.37. *Let $T_0 = \sigma_1 + \sigma_2$. Then, if $T < T_0$, system (4.99) admits non-trivial solutions $f_i \in C^\infty([0, T])$, $i = 1, 2$.*

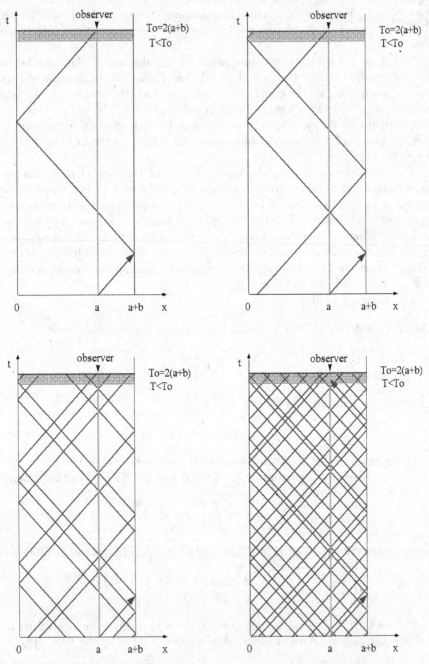

Fig. 4.5. Construction of the support of a non-observable solution

Before proving this lemma let us see how Theorem 4.36 may be obtained from it. It is clear that it is sufficient to prove Theorem 4.36 for large values of T so that we assume, without loss of generality, that $T \geq 2(\ell_0 + \hat{\ell})$, where $\hat{\ell} = \max(\ell_1, \ell_2)$.

Let f_1, f_2 be non-zero solutions of (4.99) for $\sigma_1 = 2\ell_1$, $\sigma_2 = 2\ell_2$ and $\tilde{T} = T - 2\ell_0$. We define the functions

$$\phi^i(t, x) = \frac{1}{2} \int_{t-x}^{t+x} f_i(\tau - \ell_0) d\tau, \qquad i = 1, 2,$$

for $x \in [0, \ell_i]$, $t \in [x + \ell_0, T - \ell_0 - x]$. These functions satisfy

$$(S_i) \qquad \begin{cases} \phi_{tt}^i(t, x) = \phi_{xx}^i(t, x) \\ \phi^i(t, 0) = \phi^i(t, \ell_i) = 0, \end{cases}$$

whenever $x \in [0, \ell_i]$ and $t \in [\ell_0 + x, T - \ell_0 - x]$.

Each of the functions ϕ^i may be extended to a solution of (S_i), which we will denote again by ϕ^i, defined in the region $[\ell_0, T - \ell_0] \times [0, \ell_i]$. Note that these functions have been chosen such that $\phi_x^i(t, 0) = f_i(t)$ for $t \in [\ell_0, T - \ell_0]$. Besides,

$$\phi_x^1(t, 0) + \phi_x^2(t, 0) = f_1(t) + f_2(t) = 0 \quad \text{and} \quad \phi^1(t, 0) = \phi^2(t, 0) = 0.$$

Then, $\bar{\phi} = (\phi^0 = 0, \phi^1, \phi^2)$ is a solution of (4.2) defined in the time interval $[\ell_0, T - \ell_0]$. Consequently, the unique solution of (4.2) defined on $[0, T]$ that coincides with $\bar{\phi}$ on $[\ell_0, T - \ell_0]$ satisfies the vanishing condition (4.98).

It just remains to prove that the initial data of $\bar{\phi}$ belong to $\mathcal{W}^\alpha$ for every real α.

As $\phi^0 \equiv 0$ and $f_1, f_2 \in C^\infty([0, T])$ this is equivalent to proving that for some $T^* \in [\ell_0, T - \ell_0]$ and every $k \in \mathbb{N}$ the following inequalities hold

$$\frac{\partial^{2k}}{\partial x^{2k}} \phi^i(T^*, 0) = \frac{\partial^{2k}}{\partial x^{2k}} \phi^i(T^*, \ell_i) = \frac{\partial^{2k+1}}{\partial x^{2k+1}} \phi^1(T^*, 0) + \frac{\partial^{2k+1}}{\partial x^{2k+1}} \phi^2(T^*, 0) = 0,$$

$$\tag{4.100}$$

and

$$\frac{\partial^{2k}}{\partial x^{2k}} \phi_t^i(T^*, 0) = \frac{\partial^{2k}}{\partial x^{2k}} \phi_t^i(T^*, \ell_i) = \frac{\partial^{2k+1}}{\partial x^{2k+1}} \phi_t^1(T^*, 0) + \frac{\partial^{2k+1}}{\partial x^{2k+1}} \phi_t^2(T^*, 0) = 0,$$

$$\tag{4.101}$$

for $i = 1, 2$. These facts guarantee and characterize the property that the data of the constructed solution $\bar{\phi}$ belong to any power of the domain of the generator of the three-string system.

To check these identities we first observe that, if f is a smooth function, then

$$(I_\sigma f)^{(k)}(t) = (I_\sigma f^{(k)})(t).$$

This implies that, if f_1 and f_2 are smooth solutions of (4.99) then so are the functions $f_1^{(k)}$, $f_2^{(k)}$ for all $k \geq 1$.

Note also that, since ϕ^i solve the wave equation, in (4.100) and (4.101) one can replace any derivative of even order in x by the derivative of the same order in t.

Combining these facts and choosing, e.g., $T^* = \ell_0 + \hat{\ell}$ we obtain the equalities (4.100) and (4.101).

The proof of Lemma 4.37 is based on the following facts:

Proposition 4.38. *If $\sigma_1/\sigma_2 \in \mathbb{Q}$ then, for every $T > 0$ there exist non-trivial functions $\varphi \in C^\infty([0,T])$ such that*

$$I_{\sigma_1}\varphi = 0 \quad in \quad [0, T - \sigma_1], \quad I_{\sigma_2}\varphi = 0 \quad in \quad [0, T - \sigma_2]. \tag{4.102}$$

Proof. If $\sigma_1/\sigma_2 \in \mathbb{Q}$ there exist numbers $p, q \in \mathbb{N}$, $\gamma \in \mathbb{R}$ such that

$$\frac{\sigma_1}{p} = \frac{\sigma_2}{q} = \gamma.$$

Let $\varphi \in C^\infty(\mathbb{R})$ be a non trivial, γ-periodic function such that

$$\int_0^\gamma \varphi(\tau) d\tau = 0.$$

Then,

$$I_{\sigma_1}\varphi = \int_t^{t+\sigma_1} \varphi(\tau) d\tau = \int_t^{t+\gamma p} \varphi(\tau) d\tau = p \int_t^{t+\gamma} \varphi(\tau) d\tau = 0.$$

In a similar way it may be proved that $I_{\sigma_2}\varphi = 0$.

Proposition 4.39. *Let $\varepsilon > 0$, $T = \sigma_1 + \sigma_2 - \varepsilon$ and $\sigma_1/\sigma_2 \notin \mathbb{Q}$. Then, there exists a non-zero function $\varphi \in C^\infty([0,T])$, such that (4.102) holds.*

Proof. The real number σ_2 may be expressed as $\sigma_2 = n\sigma_1 + \omega$, $n \in \mathbb{N}$, $\omega \in (0, \sigma_1)$. Since σ_1/σ_2 is irrational, so are ω/σ_1 and ω/σ_2. Let us consider the sequence $\{\omega_k\}$ defined by

$$\omega_k \in (0, \sigma_1), \quad k\omega - \omega_k \in \sigma_1 \mathbb{Z}$$

(the values of $k\omega$ modulo σ_1). As a consequence of the irrationality of ω/σ_1, we have $\omega_k \neq \omega_l$ if $k \neq l$ and that the sequence $\{\omega_k\}$ is dense in the interval $[0, \sigma_1]$. Then there exist $k_1 < 0$, $k_2 > 0$ such that $\omega_{k_1}, \omega_{k_2} \in [\sigma_1 - \varepsilon, \sigma_1)$ and $\omega_k \in [0, \sigma_1 - \varepsilon)$ for every k satisfying $k_1 < k < k_2$[5].

Now let us define the subsets of $[0, \sigma_1)$:

$$\Omega_k = (\omega_k, \omega_k + \gamma)$$

---

[5] In other words, $k_1 < 0$ and $k_2 > 0$ are the values of k with the smallest non-zero absolute value such that ω_k is in the interval $[\sigma_1 - \varepsilon, \sigma_1)$.

for $k_1 < k \leq k_2$, where $\gamma > 0$ is sufficiently small so that it holds

$$\overline{\Omega_k} \cap \overline{\Omega_l} = \emptyset \text{ if } k \neq l \text{ and } \Omega_k \subset (0, \sigma_1) \text{ for } k_1 < k, l \leq k_2.$$

It is not difficult to show that the sets Ω_k have the following properties:
(i) if $t \in \Omega_k$ with $k_1 < k < k_2$ and $t = \omega_k + \tau$ for some $\tau \in (0, \gamma)$ then,

$$t + \omega = \omega_{k+1} + \tau \qquad \text{if } \omega_k < \omega_{k+1},$$
$$t + \omega = \omega_{k+1} + \tau - \sigma_1 \text{ if } \omega_{k+1} < \omega_k.$$

(ii) if $t \in [0, \sigma_1 - \varepsilon) \setminus \cup \Omega_k$, then,

$$t + \omega \notin \cup \Omega_k \qquad \text{if } t < \sigma_1 - \omega,$$
$$t + \omega - \sigma_1 \notin \cup \Omega_k \text{ if } t \geq \sigma_1 - \omega.$$

Let us choose now a function $\psi \in C^\infty([0, \sigma_1])$ with support contained in the interval $(0, \gamma)$ and satisfying $\int_0^\gamma \psi(\tau) d\tau = 0$ and define the function φ in $[0, \sigma_1]$ by

$$\varphi(t) = \begin{cases} \varphi(t - \omega_k) \text{ if } t \in \Omega_k, \\ 0 \qquad \text{if } t \in [0, \sigma_1] \setminus \cup \Omega_k. \end{cases}$$

Then it follows $\varphi \in C^\infty([0, \sigma_1])$ and $\operatorname{supp} \varphi \subset \cup \Omega_k \subset (0, \sigma_1)$. In particular, the σ_1-periodic extension of φ to $\mathbb{R}$, which we still denote by φ, verifies $\varphi \in C^\infty(\mathbb{R})$.

Now, let us check that φ is in addition one of the functions, whose existence is asserted in the Proposition.

Let $t_1, t_2 \in [0, \sigma_1] \setminus \cup \Omega_k$, then

$$\int_{t_1}^{t_2} \varphi(\tau) d\tau = \sum_{m: \Omega_m \subset (t_1, t_2)} \int_{\Omega_m} \varphi(\tau) d\tau = \sum_m \int_{\omega_m}^{\omega_m + \gamma} \varphi(\tau - \omega_m) d\tau$$

$$= \sum_m \int_0^\gamma \varphi(\tau) d\tau = 0.$$

In particular, if we choose $t_1 = 0$, $t_2 = \sigma_1$ we get $\int_0^{\sigma_1} \varphi(\tau) d\tau = 0$, and therefore, since φ is σ_1-periodic, $I_{\sigma_1} \varphi = 0$ in $\mathbb{R}$.

It remains to calculate $I_{\sigma_2} \varphi$ for the values of t in the interval $[0, \sigma_1 - \varepsilon)$. Two cases are possible:

Case 1. $t \in \Omega_k$ for some k. Then, $t = \omega_k + \tau$ with $\tau \in (0, \gamma)$. We will assume that $\omega_k < \omega_{k+1}$, since when $\omega_{k+1} < \omega_k$ the result is obtained in a similar way.

In view of the property (i) of the sets Ω_k mentioned above, we obtain

$$I_{\sigma_2} \varphi(t) = \int_t^{t+\omega} \varphi(s) ds = \int_{\omega_k + \tau}^{\omega_{k+1} + \tau} \varphi(s) ds$$

$$= \int_{\omega_k + \tau}^{\omega_k + \gamma} \varphi(s) ds + \int_{\omega_{k+1}}^{\omega_{k+1} + \tau} \varphi(s) ds + \int_{\omega_k + \gamma}^{\omega_{k+1}} \varphi(s) ds.$$

But $\omega_k + \gamma$ and ω_{k+1} do not belong to $\cup\Omega_k$, $\int_{\omega_k+\gamma}^{\omega_{k+1}}\varphi(s)ds = 0$, and thus

$$I_{\sigma_2}\varphi(t) = \int_{\omega_k+\tau}^{\omega_k+\gamma}\varphi(s - \omega_k)ds + \int_{\omega_{k+1}}^{\omega_{k+1}+\tau}\varphi(s - \omega_{k+1})ds$$

$$= \int_\tau^\gamma \psi(s)ds + \int_0^\tau \psi(s)ds = 0.$$

Case 2. $t \notin \cup\Omega_k$. In view of the property (ii), if $t < \sigma_1 - \omega$, then $t + \omega$ does not belong to $\cup\Omega_k$ and we have

$$I_{\sigma_2}\varphi(t) = \int_t^{t+\omega}\varphi(s)ds = 0.$$

If $t \geq \sigma_1 - \omega$, then $t - \sigma_1 + \omega \notin \cup\Omega_k$ and it holds

$$I_{\sigma_2}\varphi(t) = \int_t^{t+\omega-\sigma_1}\varphi(s)ds = 0.$$

This proves the Proposition.

Proof (Proof of Lemma 4.37). It follows immediately from the previous propositions. It suffices to take $f_1 = \varphi$ and $f_2 = -\varphi$ according to Propositions 4.38 or 4.39, depending on whether σ_1/σ_2 is rational or irrational.

4.10 Application of the Method of Moments to the Control of the Three String Network

In this section we study the problem of moments (3.40), i. e.

$$\int_0^T \varkappa_{|k|}e^{i\lambda_k t}\, h(t)dt = u_{1,|k|} - i\lambda_k u_{0,|k|} \quad \text{for every } k \in \mathbb{Z}_*, \qquad (4.103)$$

for the three string network. Recall that, in view of the results of Section 3.3, the existence of a solution $h \in L^2(0,T)$ of the problem of moments (4.103) is equivalent to the controllability in time T of the initial state $(\bar{u}_0, \bar{u}_1)$ with

$$\bar{u}_0 = \sum_{n\in\mathbb{N}} u_{0,n}\bar{\theta}_n, \qquad \bar{u}_1 = \sum_{n\in\mathbb{N}} u_{1,n}\bar{\theta}_n.$$

Thus, this is an alternative way to study the control problem for system (4.33).

Performing the change of variable $t \to t - T/2$ in (4.103), we obtain

$$\int_{-\frac{T}{2}}^{\frac{T}{2}} e^{i\lambda_n t}\, h(t - \frac{T}{2})dt = \frac{1}{\varkappa_n}\left(u_{1,n} - i\lambda_n u_{0,n}\right)e^{i\lambda_n T/2}.$$

Denoting $m_n := \varkappa_n^{-1} (u_{1,n} - i\lambda_n u_{0,n}) e^{i\lambda_n T/2}$, $A := T/2$, problem (4.103) will be written in the form (3.28). This implies, in account of Proposition 3.18, that, if we can construct a sequence (v_n) biorthogonal to $(e^{i\lambda_n t})$ in $L^2(-T/2, T/2)$ then the initial states satisfying

$$\sum_{n \in \mathbb{Z}_*} \left| \frac{1}{\varkappa_n} (u_{1,n} - i\lambda_n u_{0,n}) e^{i\lambda_n T/2} \right| \|v_n\|_{L^2} < \infty \qquad (4.104)$$

are controllable in time T with control

$$v = \sum_{n \in \mathbb{Z}_*} \frac{1}{\varkappa_n} (u_{1,n} - i\lambda_n u_{0,n}) e^{i\lambda_n T/2} v_n.$$

Inequality (4.104) is equivalent to

$$\sum_{n \in \mathbb{Z}_*} \frac{1}{|\varkappa_n|} (|u_{1,n}| + \lambda_n |u_{0,n}|) \|v_n\|_{L^2} < \infty.$$

In particular, if the biorthogonal sequence (v_n) has been obtained from a generating function F then all the initial states satisfying

$$\sum_{n \in \mathbb{Z}_*} \frac{1}{|\varkappa_n| |F'(\lambda_n)|} (|u_{1,n}| + \lambda_n |u_{0,n}|) < \infty \qquad (4.105)$$

are controllable in time T.

Let us remark that for the three string network it is easy to construct a generating function, since we already know a function that vanishes at the numbers λ_n. Indeed, recall that, as it has been shown in Proposition 4.20, $q(\lambda_n) = 0$, where

$$q(z) = \cos z\ell_0 \sin z\ell_1 \sin z\ell_2 + \sin z\ell_0 \cos z\ell_1 \sin z\ell_2$$
$$+ \sin z\ell_0 \cos z\ell_1 \sin z\ell_1 \cos z\ell_2, \qquad (4.106)$$

and this is an entire function bounded on the real axis: $|q(z)| \le 3$.

On the other hand, if we replace in (4.106) $\cos(z\ell_k)$ and $\sin(z\ell_k)$ by their expressions in terms of complex exponentials

$$\cos z\ell_k = \frac{1}{2} \left(e^{iz\ell_k} + e^{-iz\ell_k} \right), \qquad \sin z\ell_k = -\frac{i}{2} \left(e^{iz\ell_k} - e^{-iz\ell_k} \right),$$

we see that q may be written as a linear combination of eight terms of the form e^{izh}, with

$$|h| \le \ell_0 + \ell_1 + \ell_2.$$

Then, there exists a constant $C > 0$ such that, for every $z \in \mathbb{C}$,

$$|q(z)| \le C e^{|z|(\ell_0 + \ell_1 + \ell_2)},$$

that is, the function q is of exponential type at most $\ell_0 + \ell_1 + \ell_2$.

Then, based on the results of Subsection 3.3.1, we can assert that there exists a sequence (v_n) biorthogonal to $(e^{i\lambda_n t})$ in any interval $(-T/2, T/2)$ with $T \geq 2(\ell_0 + \ell_1 + \ell_2)$ that satisfies

$$\|v_n\|_{L^2(-\frac{T}{2}, \frac{T}{2})} \leq \frac{C}{|q'(\lambda_n)|}, \quad n \in \mathbb{N}, \tag{4.107}$$

where the constant $C > 0$ does not depend on n.

This guarantees immediately that the spaces of sequences for which the problem of moments (4.103) has a solution is dense in l^2. Therefore, the space of controllable initial states in time $T \geq 2(\ell_0 + \ell_1 + \ell_2)$, is dense in $H \times V'$. Moreover, all the initial states from $Z \times Z$ are controllable. That means that spectral controllability holds as well.

Now we estimate $|q'(\lambda_n)|$ in order to identify larger subspaces of controllable initial states. Observe that the function q may be written in the form

$$q(z) = \sin z\ell_0 \sin z\ell_1 \sin z\ell_2 \left(\cot z\ell_0 + \cot z\ell_1 + \cot z\ell_2\right). \tag{4.108}$$

Then

$$|q'(\lambda_n)| = |\sin \lambda_n \ell_0 \sin \lambda_n \ell_1 \sin \lambda_n \ell_2| \, \mathbf{A}_n, \tag{4.109}$$

with

$$\mathbf{A}_n = \left(\frac{\ell_0}{\sin^2 \lambda_n \ell_0} + \frac{\ell_1}{\sin^2 \lambda_n \ell_1} + \frac{\ell_2}{\sin^2 \lambda_n \ell_2}\right).$$

In account of (4.105), we can ensure that the initial states satisfying

$$\sum_{n \in \mathbb{Z}_*} \frac{1}{|\varkappa_n| \, |q'(\lambda_n)|} \left(|u_{1,n}| + \lambda_n |u_{0,n}|\right) < \infty \tag{4.110}$$

are controllable.

To make this condition more precise, we need to estimate the product $|\varkappa_n| \, |q'(\lambda_n)|$. Recall that (see Remark 4.31)

$$|\varkappa_n| = \frac{\sqrt{2}\lambda_n}{|\sin \lambda_n \ell_0|} \mathbf{A}_n^{-\frac{1}{2}},$$

and thus we have

$$|\varkappa_n| \, |q'(\lambda_n)| = \sqrt{2}\lambda_n \, |\sin \lambda_n \ell_1 \sin \lambda_n \ell_2| \, \mathbf{A}_n^{\frac{1}{2}}.$$

Then,

$$|\varkappa_n|^2 \, |q'(\lambda_n)|^2 = 2\lambda_n^2 \, |\sin \lambda_n \ell_1 \sin \lambda_n \ell_2|^2 \left(\frac{\ell_0}{\sin^2 \lambda_n \ell_0} + \frac{\ell_1}{\sin^2 \lambda_n \ell_1} + \frac{\ell_2}{\sin^2 \lambda_n \ell_2}\right)$$

$$\geq 2\lambda_n^2 \left(\ell_1 \sin^2 \lambda_n \ell_2 + \ell_2 \sin^2 \lambda_n \ell_1\right) \geq C\lambda_n^2 c_n^2.$$

Here $c_n = \max\left(|\sin \lambda_n \ell_1|, |\sin \lambda_n \ell_2|\right)$ are the coefficients defined by (4.75) in Section 4.6.

With this we can conclude that a sufficient condition for the initial state $(\bar{u}_0, \bar{u}_1)$ to be controllable is

$$\sum_{n \in \mathbb{Z}_*} \frac{1}{c_n \lambda_n} \left(|u_{1,n}| + \lambda_n |u_{0,n}|\right) < \infty. \tag{4.111}$$

Let us observe that this result is weaker than that given in Proposition 4.6, since, if the initial state $(\bar{u}_0, \bar{u}_1)$ satisfies (4.111) then it also satisfies

$$\sum_{n \in \mathbb{Z}_*} \frac{1}{c_n^2 \lambda_n^2} \left(u_{1,n}^2 + \lambda_n u_{0,n}^2\right) < \infty.$$

Let us choose $\delta > 0$. The series

$$\sum_{n \in \mathbb{Z}_*} \frac{1}{\lambda_n^{1+\delta}}$$

converges for every $\delta > 0$, as

$$\lim_{n \to \infty} \frac{\lambda_n}{n} = \frac{\pi}{\ell_0 + \ell_1 + \ell_2}.$$

Then, with the help of the Cauchy-Schwarz inequality we obtain

$$\sum_{n \in \mathbb{Z}_*} \frac{1}{c_n \lambda_n} \left(|u_{1,n}| + \lambda_n |u_{0,n}|\right)$$

$$\leq \left(\sum_{n \in \mathbb{Z}_*} \frac{\lambda_n^{\delta-1}}{c_n^2} \left(u_{1,n}^2 + \lambda_n^2 u_{0,n}^2\right)\right)^{1/2} \left(\sum_{n \in \mathbb{Z}_*} \frac{1}{\lambda_n^{1+\delta}}\right)^{1/2}$$

$$\leq C \left(\sum_{n \in \mathbb{Z}_*} \frac{\lambda_n^{\delta-1}}{c_n^2} \left(u_{1,n}^2 + \lambda_n^2 u_{0,n}^2\right)\right)^{1/2}.$$

Thus, for (4.111) to be verified and consequently for the initial state $(\bar{u}_0, \bar{u}_1)$ to be controllable in time T, it is sufficient that

$$\sum_{n \in \mathbb{Z}_*} \frac{\lambda_n^{\delta-1}}{c_n^2} \left(u_{1,n}^2 + \lambda_n^2 u_{0,n}^2\right) < \infty.$$

In particular, if $\ell_1/\ell_2 \in \mathbf{B}_\varepsilon$, in view of (4.85) we have

$$c_n \geq \frac{C}{\lambda_n^{1+\varepsilon}},$$

and, consequently, the controllability condition (4.111) obtained with the method of moments guarantees that all the initial states from

$$(\bar{u}_0, \bar{u}_1) \in \mathcal{W}^{\frac{3}{2}+\varepsilon+\delta} = V^{\frac{3}{2}+\varepsilon+\delta} \times V^{\frac{1}{2}+\varepsilon+\delta},$$

with arbitrarily small $\delta > 0$ are controllable.

Thus, we need roughly $1/2$ more derivatives in L^2 on the initial data for the method of moments than in Corollary 4.30. This difference may be due to the possible inaccuracy in the estimates of the sequence $|\varkappa_n| |q'(\lambda_n)|$.

Remark 4.40. According to Proposition 3.25, once we have identified subspaces of controllable initial states in time T of the form $\mathcal{W}^r$, we can construct *a posteriori* a sequence $(\tilde{v}_n)$, biorthogonal to $(e^{i\lambda_n t})$ in $L^2(0,T)$, satisfying

$$\|\tilde{v}_n\|_{L^2(0,T)} \le C\lambda_n^{r-1}.$$

Thus, in view of Corollary 4.30, if $\ell_1/\ell_2 \in \mathbf{B}_\varepsilon$, for the system of the three string network, a sequence $(\tilde{v}_n)$ biorthogonal to $(e^{i\lambda_n t})$ in $L^2(0,T)$ can be constructed verifying

$$\|\tilde{v}_n\|_{L^2(0,T)} \le C\lambda_n^\varepsilon. \tag{4.112}$$

Let us remark that the biorthogonal sequence (v_n) used in this section does not necessarily coincide with $(\tilde{v}_n)$. Recall in addition, that we do not resort to that sequence, since we got information on controllable subspaces without using the information provided by Corollary 4.30.

Let us try a sharper estimate of the sequence (v_n). In view of (4.107) it suffices to estimate $|q'(\lambda_n)|$. From equality (4.108) we obtain

$$|q'(\lambda_n)| \ge \ell_0 |\sin \lambda_n \ell_1 \sin \lambda_n \ell_2| + \ell_1 |\sin \lambda_n \ell_0 \sin \lambda_n \ell_2| + \ell_2 |\sin \lambda_n \ell_0 \sin \lambda_n \ell_1|$$
$$\ge Cs(\lambda, \ell_0, \ell_1, \ell_2),$$

where we have denoted

$$s(\lambda, \ell_0, \ell_1, \ell_2) := |\sin \lambda_n \ell_0| |\sin \lambda_n \ell_1| + |\sin \lambda_n \ell_0| |\sin \lambda_n \ell_2| + |\sin \lambda_n \ell_1| |\sin \lambda_n \ell_2|.$$

To obtain lower bounds of the function s we need to impose additional restrictions on the lengths ℓ_0, ℓ_1, ℓ_2. Let us assume that those lengths satisfy the following rational approximation conditions, which we will call briefly *conditions* (S) (see also Definition A.9 in Appendix A):

- ℓ_0, ℓ_1, ℓ_2 are linearly independent over the field $\mathbb{Q}$ of rational numbers;
- all the ratios ℓ_i/ℓ_j are algebraic numbers, that is, roots of polynomials with rational coefficients.

Under these hypotheses in Proposition A.11 it is proved that for every $\varepsilon > 0$ there exists a constant $C_\varepsilon > 0$ such that for every $n = 1, 2, ...,$ the following inequality holds

$$s(\lambda_n, \ell_0, \ell_1, \ell_2) \ge C_\varepsilon (\lambda_n)^{-1-e}.$$

This guarantees that

$$\|v_n\|_{L^2(-T/2,T/2)} \leq C\lambda_n^{1+\varepsilon}.$$

Unfortunately, we have imposed restrictive conditions on ℓ_0 and we have been able to prove an estimate weaker than (4.112). This could be caused by two reasons: that the norms of the elements of the sequence (v_n) are actually larger than those of the elements of the sequences $(\tilde{v}_n)$ or that the technique we have used to estimate $|q'(\lambda_n)|$ is not sharp.

The obtention of the optimal controllability results for the three string network by the method of moments is therefore an open problem.

5

General Trees

In this chapter we study the control problem for networks of strings, which are supported on a tree-shaped graph, when the control acts on one exterior node. We will follow the technique described in Chapter 4 for the three string network, which is the simplest example of a network supported by a tree-shaped graph, not reduced to a single string.

Let us briefly recall this technique. The key element is the construction of an operator $\mathbf{B} : V \times H \to V \times H$, which guarantees the existence of a constant $C > 0$ such that all the solutions of the homogeneous system (2.11)-(2.16) with initial states $(\bar{\phi}_0, \bar{\phi}_1) \in Z \times Z$ satisfy the observability inequality

$$C \int_0^T |\partial_n \phi^1(t, \mathbf{v}_1)|^2 dt \geq \|\mathbf{B}(\bar{\phi}_0, \bar{\phi}_1)\|_{V \times H}^2,$$

where T is twice the total length of the network (here we used the notations introduced in Chapter 2 for general networks).

The operator $\mathbf{B}$ has the property of being essentially diagonal: there exist real numbers b_n such that

$$\|\mathbf{B}(\bar{\phi}_0, \bar{\phi}_1)\|_{V \times H}^2 = \sum_{n \in \mathbb{N}} b_n^2 \left(\mu_n \phi_{0,n}^2 + \phi_{1,n}^2 \right).$$

This leads to the inequality

$$C \int_0^T |\partial_n \phi^1(t, \mathbf{v}_1)|^2 dt \geq \sum_{n \in \mathbb{N}} b_n^2 \left(\mu_n \phi_{0,n}^2 + \phi_{1,n}^2 \right),$$

which allows to determine subspaces of $H \times V'$ of controllable initial states in time T.

5.1 Notations and Statement of the Problem

5.1.1 Notations for Graphs

In this section, we introduce precise notations for the elements of the rest configuration graph. This is needed to write the equations of the motion of the network in a way that takes into account the topological structure of the graph.

Let $\mathcal{A}$ be a planar, connected graph without closed paths. According to the usual terminology in Graph Theory, those graphs will be called *trees*. By the multiplicity of a vertex of $\mathcal{A}$ we mean the number of edges that branch out from it. If the multiplicity is equal to one, the vertex is called exterior, otherwise, it is said to be interior. We assume that the graph $\mathcal{A}$ does not contain vertices of multiplicity two, since they are irrelevant for our model.

In what follows, we describe a procedure for indexing the edges and vertices of the graph. In Figure 5.1 an example is given of a tree with indices defined according to this rule. First, we choose an exterior vertex and denote it by $\mathcal{R}$. It is called the root of $\mathcal{A}$. The remaining edges and vertices will be denoted by $\mathbf{e}_{\bar{\alpha}}$ and $\mathcal{O}_{\bar{\alpha}}$, respectively, where $\bar{\alpha} = (\alpha_1, \ldots, \alpha_k)$ is a multi-index (possibly empty) of variable length k defined by recurrence for every edge in the following way.

For the edge containing the root $\mathcal{R}$ we choose the empty index. Thus, that edge is denoted by $\mathbf{e}$ and its vertex different from $\mathcal{R}$ is denoted by $\mathcal{O}$.

Assume now that the interior vertex $\mathcal{O}_{\bar{\alpha}}$, contained in the edge $\mathbf{e}_{\bar{\alpha}}$, has multiplicity equal to $m_{\bar{\alpha}} + 1$. This means that there are $m_{\bar{\alpha}}$ edges, different from $\mathbf{e}_{\bar{\alpha}}$, that branch out from $\mathcal{O}_{\bar{\alpha}}$. We denote these edges by $\mathbf{e}_{\bar{\alpha}\circ\beta}$, $\beta = 1, \ldots, m_{\bar{\alpha}}$ and the other vertex of the edge $\mathbf{e}_{\bar{\alpha}\circ\beta}$ by $\mathcal{O}_{\bar{\alpha}\circ\beta}$. Here, $\bar{\alpha} \circ \beta$ represents the index $(\alpha_1, \ldots, \alpha_k, \beta)$, obtained by adding a new component β to the index $\bar{\alpha} = (\alpha_1, \ldots, \alpha_k)$. In general, if $\bar{\alpha} = (\alpha_1, \ldots, \alpha_k)$ and $\bar{\beta} = (\beta_1, \ldots, \beta_m)$, then $\bar{\alpha} \circ \bar{\beta}$ will denote the multi-index of length $k + m$ defined by $\bar{\alpha} \circ \bar{\beta} = (\alpha_1, \ldots, \alpha_k, \beta_1, \ldots, \beta_m)$.

Let now $\mathcal{M}$ be the set of the interior vertices of $\mathcal{A}$ and $\mathcal{S}$ the set of exterior vertices, $\mathcal{R}$ being excepted, and define

$$\mathcal{I}_{\mathcal{M}} = \{\bar{\alpha}; \quad \mathcal{O}_{\bar{\alpha}} \in \mathcal{M}\}, \qquad \mathcal{I}_{\mathcal{S}} = \{\bar{\alpha}; \quad \mathcal{O}_{\bar{\alpha}} \in \mathcal{S}\},$$

which are the sets of the indices of the interior and exterior vertices, except $\mathcal{R}$, respectively. Note that with these notations, we admit the empty multi-index, which corresponds to the vertex $\mathcal{O}$ and belongs to one of the sets $\mathcal{I}_{\mathcal{M}}$ or $\mathcal{I}_{\mathcal{S}}$. Finally, $\mathcal{I} = \mathcal{I}_{\mathcal{S}} \bigcup \mathcal{I}_{\mathcal{M}}$ is the set of the indices of all the vertices, except that of the root $\mathcal{R}$.

Further, for $\bar{\alpha} \in \mathcal{I}_{\mathcal{M}}$, the sets

$$\mathcal{A}_{\bar{\alpha}} = \{\mathbf{e}_{\bar{\alpha}\circ\bar{\beta}}; \quad \bar{\alpha} \circ \bar{\beta} \in \mathcal{I}\}$$

are called sub-trees of $\mathcal{A}$. Note that $\mathcal{A}_{\bar{\alpha}}$ is formed by the edges having indices with a common initial part $\bar{\alpha}$. This means that $\mathcal{A}_{\bar{\alpha}}$ is also a tree branching out from the vertex of $\mathbf{e}_{\bar{\alpha}}$ different from $\mathcal{O}_{\bar{\alpha}}$. Then, if one chooses that vertex as the root $\mathcal{R}_{\bar{\alpha}}$ of $\mathcal{A}_{\bar{\alpha}}$ and denotes by $\mathbf{e}_{\bar{\beta}}^{\bar{\alpha}}$ the edge with index $\bar{\beta}$ in $\mathcal{A}_{\bar{\alpha}}$, according to the numbering rule defined above for trees, it holds that

$$\mathbf{e}_{\bar{\alpha}\circ\bar{\beta}} = \mathbf{e}_{\bar{\beta}}^{\bar{\alpha}} \qquad \mathcal{O}_{\bar{\alpha}\circ\bar{\beta}} = \mathcal{O}_{\bar{\beta}}^{\bar{\alpha}}.$$

In order to prove properties of trees, we shall often proceed by induction with respect to the largest length of the indices $\bar{\alpha}$ used to number the edges according to the procedure described above. To do this we should prove that:

1. The property is true for the simplest case of a one-edged tree (i.e., the corresponding network is formed by a single string).
2. If the property is true for all the sub-trees $\mathcal{A}_1, \ldots, \mathcal{A}_m$ branching out from $\mathcal{O}$, then it is also true for the whole tree $\mathcal{A}$.

In what follows such process will be called simply induction.

Besides, the length of the edge $\mathbf{e}_{\bar{\alpha}}$ will be denoted by $\ell_{\bar{\alpha}}$. Then, $\mathbf{e}_{\bar{\alpha}}$ may be parameterized by its arc length by means of the functions $\pi_{\bar{\alpha}}$, defined in $[0, \ell_{\bar{\alpha}}]$ such that $\pi_{\bar{\alpha}}(\ell_{\bar{\alpha}}) = \mathcal{O}_{\bar{\alpha}}$ and $\pi_{\bar{\alpha}}(0)$ is the other vertex of this edge.

Finally, we denote by L_A and $L_{\bar{\alpha}}$, $\bar{\alpha} \in \mathcal{I}$, the sum of the lengths of all the edges of the tree $\mathcal{A}$ (i.e., the total length of $\mathcal{A}$) and of its sub-trees $\mathcal{A}_{\bar{\alpha}}$, respectively.

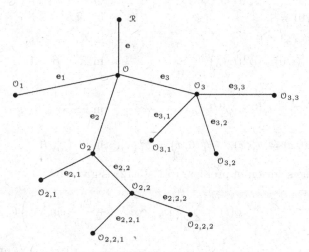

Fig. 5.1. A tree with indices for its vertices and edges

5.1.2 Equations of Motion

In this subsection we write the equations of the motion of the tree-shaped network with a controlled node (2.11)-(2.16) with the specific notations introduced for trees in this chapter. The vertex of the graph $\mathcal{A}$ corresponding to the controlled node of the network has been chosen as the root of the tree. The system of equations reads as follows:

$$u_{tt}^{\bar{\alpha}}(t,x) = u_{xx}^{\bar{\alpha}}(t,x) \qquad\qquad \text{in } \mathbb{R} \times [0,\ell_{\bar{\alpha}}], \quad \bar{\alpha} \in \mathcal{I}, \qquad (5.1)$$

$$u(t,0) = v(t), \qquad\qquad\qquad \text{in } \mathbb{R}, \qquad\qquad\qquad (5.2)$$

$$u^{\bar{\alpha}}(t,\ell_{\bar{\alpha}}) = 0 \qquad\qquad\qquad \text{in } \mathbb{R}, \ \ \bar{\alpha} \in \mathcal{I}_{\mathcal{S}}, \qquad\qquad (5.3)$$

$$u^{\bar{\alpha}\circ\beta}(t,0) = u^{\bar{\alpha}}(t,\ell_{\bar{\alpha}}) \qquad \text{in } \mathbb{R}, \ \beta = 1,\ldots,m_{\bar{\alpha}}, \bar{\alpha} \in \mathcal{I}_{\mathcal{M}}, \quad (5.4)$$

$$\sum_{\beta=1}^{m_{\bar{\alpha}}} u_x^{\bar{\alpha}\circ\beta}(t,0) = u_x^{\bar{\alpha}}(t,\ell_{\bar{\alpha}}) \qquad \text{in } \mathbb{R}, \ \ \bar{\alpha} \in \mathcal{I}_{\mathcal{M}}, \qquad\qquad (5.5)$$

$$u^{\bar{\alpha}}(0,x) = u_0^{\bar{\alpha}}(x), \quad u_t^{\bar{\alpha}}(0,x) = u_1^{\bar{\alpha}}(x), \quad \text{in } [0,\ell_{\bar{\alpha}}], \ \bar{\alpha} \in \mathcal{I}. \quad (5.6)$$

For every $\bar{\alpha} \in \mathcal{I}$, the function $u^{\bar{\alpha}}(t,x) : \mathbb{R} \times [0,\ell_{\bar{\alpha}}] \to \mathbb{R}$ denotes the transversal displacement of the string with index $\bar{\alpha}$. We will denote by $\bar{u}$ the set whose elements are $u^{\bar{\alpha}}$, $\bar{\alpha} \in \mathcal{I}$. In particular, the sets of initial states $(u_0^{\bar{\alpha}})_{\bar{\alpha}\in\mathcal{I}}, (u_1^{\bar{\alpha}})_{\bar{\alpha}\in\mathcal{I}}$ of the strings are denoted by $\bar{u}^0$ and $\bar{u}^1$. With these notations, the remaining elements relative to system (5.1)-(5.6) are defined exactly as in Subsection 2.2.2 of Chapter 2.

We also consider the homogeneous version of system (5.1)-(5.6):

$$\phi_{tt}^{\bar{\alpha}}(t,x) = \phi_{xx}^{\bar{\alpha}}(t,x) \qquad\qquad \text{in } \mathbb{R} \times [0,\ell_{\bar{\alpha}}], \quad \bar{\alpha} \in \mathcal{I}, \qquad (5.7)$$

$$\phi(t,0) = 0, \qquad\qquad\qquad\qquad \text{in } \mathbb{R}, \qquad\qquad\qquad\qquad (5.8)$$

$$\phi^{\bar{\alpha}}(t,\ell_{\bar{\alpha}}) = 0 \qquad\qquad\qquad \text{in } \mathbb{R}, \ \ \bar{\alpha} \in \mathcal{I}_{\mathcal{S}}, \qquad\qquad (5.9)$$

$$\phi^{\bar{\alpha}\circ\beta}(t,0) = \phi^{\bar{\alpha}}(t,\ell_{\bar{\alpha}}) \qquad \text{in } \mathbb{R}, \quad \beta = 1, \ \ldots, m_{\bar{\alpha}}, \ \ \bar{\alpha} \in \mathcal{I}_{\mathcal{M}}, \qquad\qquad\qquad\qquad\qquad\qquad (5.10)$$

$$\sum_{\beta=1}^{m_{\bar{\alpha}}} \phi_x^{\bar{\alpha}\circ\beta}(t,0) = \phi_x^{\bar{\alpha}}(t,\ell_{\bar{\alpha}}) \qquad \text{in } \mathbb{R}, \ \ \bar{\alpha} \in \mathcal{I}_{\mathcal{M}}, \qquad\qquad (5.11)$$

$$\phi^{\bar{\alpha}}(0,x) = \phi_0^{\bar{\alpha}}(x), \quad \phi_t^{\bar{\alpha}}(0,x) = \phi_1^{\bar{\alpha}}(x), \quad \text{in } [0,\ell_{\bar{\alpha}}], \ \bar{\alpha} \in \mathcal{I}. \quad (5.12)$$

The solution of problem (5.7)-(5.12) is given by

$$\bar{\phi}(t) = \sum_{k\in\mathbb{N}} \left(\phi_{0,k} \cos \lambda_k t + \frac{\phi_{1,k}}{\lambda_k} \sin \lambda_k t \right) \bar{\theta}_k, \qquad (5.13)$$

where $(\phi_{0,k})_{k\in\mathbb{N}}, (\phi_{1,k})_{k\in\mathbb{N}}$ are the sequences of Fourier coefficients of the initial state $(\bar{\phi}^0, \bar{\phi}^1)$ in the orthonormal basis $(\bar{\theta}_k)_{k\in\mathbb{N}}$ formed by the eigenfunctions of the elliptic operator $-\Delta_A$ corresponding to (5.1)-(5.5). Recall that $(\mu_k)_{k\in\mathbb{N}}$ is the increasing sequence of eigenvalues and $\lambda_k := \sqrt{\mu_k}$.

For technical reasons, we will also consider solutions $\bar{\phi}$ of (5.7) such that $\phi^{\bar{\alpha}} \in C^2(\mathbb{R} \times [0,\ell_{\bar{\alpha}}])$, satisfying (5.9), (5.10) and (5.11), but not necessarily (5.8). That is, $\bar{\phi}$ is a smooth solution that satisfies the boundary conditions given in (5.7)-(5.11) at all the nodes, except at the root $\mathcal{R}$. These solutions will be briefly referred as *solutions of* (N). In the same way we define a *solution of* (N) *on the sub-tree* $\mathcal{A}_{\bar{\alpha}}$.

For a solution $\bar{\phi}$ of (N) we define the functions

$$G_{\bar{\alpha}}(t) := \phi_t^{\bar{\alpha}}(t,0), \qquad F_{\bar{\alpha}}(t) := \phi_x^{\bar{\alpha}}(t,0), \tag{5.14}$$

$$\widehat{G}_{\bar{\alpha}}(t) := \phi_t^{\bar{\alpha}}(t,\ell_{\bar{\alpha}}), \qquad \widehat{F}_{\bar{\alpha}}(t) := \phi_x^{\bar{\alpha}}(t,\ell_{\bar{\alpha}}), \tag{5.15}$$

for every $\bar{\alpha} \in \mathcal{I}$. These functions are the velocity and the tension at the extremes of the string $\mathbf{e}_{\bar{\alpha}}$.

According to the coupling conditions (5.10)-(5.11), we will have the formulas

$$G_{\bar{\alpha}\circ\beta}(t) = \widehat{G}_{\bar{\alpha}}(t), \qquad \sum_{\beta=1}^{m_{\bar{\alpha}}} F_{\bar{\alpha}\circ\beta}(t) = \widehat{F}_{\bar{\alpha}}(t), \tag{5.16}$$

for every $t \in \mathbb{R}$, $\bar{\alpha} \in \mathcal{I}_{\mathrm{M}}$, $\beta = 1, ..., m_{\bar{\alpha}}$.

On the other hand, from the D'Alembert formulas (3.5) we have the identities

$$\widehat{F}_{\bar{\alpha}} = \ell_{\bar{\alpha}}^+ F_{\bar{\alpha}} + \ell_{\bar{\alpha}}^- G_{\bar{\alpha}}, \qquad \widehat{G}_{\bar{\alpha}} = \ell_{\bar{\alpha}}^- F_{\bar{\alpha}} + \ell_{\bar{\alpha}}^+ G_{\bar{\alpha}},$$

for all $\bar{\alpha} \in \mathcal{I}$. In view of them, the coupling conditions (5.16) at the interior nodes may be expressed as

$$G_{\bar{\alpha}\circ\beta}(t) = \ell_{\bar{\alpha}}^- F_{\bar{\alpha}} + \ell_{\bar{\alpha}}^+ G_{\bar{\alpha}}, \tag{5.17}$$

$$\sum_{\beta=1}^{m_{\bar{\alpha}}} F_{\bar{\alpha}\circ\beta}(t) = \ell_{\bar{\alpha}}^+ F_{\bar{\alpha}} + \ell_{\bar{\alpha}}^- G_{\bar{\alpha}}. \tag{5.18}$$

For a function $\bar{w}(t)$ defined on the tree $\mathcal{A}$ the energy of $\bar{w}$ on the string $\mathbf{e}_{\bar{\alpha}}$ is defined by

$$E_{\bar{w}}^{\bar{\alpha}}(t) := \frac{1}{2} \int_0^{\ell_{\bar{\alpha}}} \left(|w_t^{\bar{\alpha}}(t,x)|^2 + |w_x^{\bar{\alpha}}(t,x)|^2 \right) dx.$$

For a sub-tree $\mathcal{A}_{\bar{\alpha}}$, we denote by $\mathbf{E}_{\bar{w}}^{\bar{\alpha}}$ the total energy of $\bar{w}$ on the sub-tree:

$$\mathbf{E}_{\bar{w}}^{\bar{\alpha}}(t) := \sum_{\beta:\bar{\alpha}\circ\bar{\beta}\in\mathcal{I}} E_{\bar{w}}^{\bar{\alpha}\circ\bar{\beta}}(t).$$

In particular, the total energy of $\bar{w}$ on the network is

$$\mathbf{E}_{\bar{w}}(t) := \sum_{\bar{\alpha}\in\mathcal{I}} E_{\bar{w}}^{\bar{\alpha}}(t).$$

5.2 The Operators $\mathcal{P}$ and $\mathcal{Q}$

In this section we define two linear operators $\mathcal{P}$ and $\mathcal{Q}$ that allow to express the relation

$$\mathcal{P}G + \mathcal{Q}F = 0 \qquad (5.19)$$

between the velocity and the tension of the solutions of (N) at the root of the tree. These operators will play an essential role in the proof of the main observability results, so we study them in detail. In particular, we need information on how they act on the traces $F_{\bar{\alpha}}$ and $G_{\bar{\alpha}}$ of the other components of the solution at the interior nodes.

First, $\mathcal{P}$ and $\mathcal{Q}$ are constructed for a string. Then, using a recursive argument, they are obtained for general trees.

5.2.1 A Tree Formed by a Single String

Assume that $\phi \in C^2(\mathbb{R} \times [0, \ell])$ satisfies the wave equation

$$\phi_{tt} - \phi_{xx} = 0$$

in $\mathbb{R} \times [0, \ell]$ and that $\phi(t, \ell) \equiv 0$. Thus, ϕ is a solution of (N) for the network formed by a single string of length ℓ. Let us note that, in this case, with the notations (5.14)-(5.15), we get $\tilde{G}(t) := \phi_t(t, \ell) = 0$.

From the D'Alembert formula (3.5) it holds

$$0 = \ell^+ G + \ell^- F, \qquad (5.20)$$

for every $t \in \mathbb{R}$. This is a relation of type (5.19) with $\mathcal{P} = \ell^+$, $\mathcal{Q} = \ell^-$.

5.2.2 Operators of Type S

We are interested not only in the existence of the operators $\mathcal{P}$ and $\mathcal{Q}$ satisfying (5.19), but also in their structure. That is why we consider a class of linear operators constituted by linear combinations of certain shift operators. This class of operators allows to describe the main properties of the operators $\mathcal{P}$ and $\mathcal{Q}$ we use in this chapter.

For the real number h we denote by τ_h the shift operator defined by

$$\tau_h f(t) := f(t + h).$$

As we shall be concerned only with algebraic properties of those operators, we may assume τ_h to act on the vector spaces of mappings $f = f(t) : \mathbb{R} \to \mathbf{W}$, where $\mathbf{W}$ is a vector space.

Let $\Lambda = \{\ell_1, \ldots, \ell_n\}$ be a set of positive numbers, not necessarily different. In what follows, whenever a set is denoted by Λ we tacitly assume that it may contain repeated elements. If $\tilde{\Lambda} = \left\{ \tilde{\ell}_1, \ldots, \tilde{\ell}_{n'} \right\}$ is another such set, we use the notation $\Lambda \sqcup \tilde{\Lambda}$ for the set $\{\ell_1, \ldots, \ell_n, \tilde{\ell}_1, \ldots, \tilde{\ell}_{n'}\}$, which once again may contain repeated elements. Observe that this operation differs from the usual union of sets in the fact that the multiplicity of the elements is taken into account.

We set

$$S(\Lambda) := span\{\tau_h : \quad h \in \mathcal{H}_\Lambda\},$$

(the set all linear combinations of shift operators τ_h with $h \in \mathcal{H}_\Lambda$) where

$$\mathcal{H}_\Lambda = \left\{ h = \sum_{i=1}^{n} \varepsilon_i \ell_i, \ \varepsilon_i = \pm 1 \right\}.$$

Observe that the set $\mathcal{H}_\Lambda$ contains at most 2^n elements, so $S(\Lambda)$ is of finite dimension.

For an operator $\mathcal{B} \in S(\Lambda)$ we shall write $s(\mathcal{B}) := s(\Lambda) := \sum_{i=1}^{n} \ell_i$. We say that $\mathcal{B}$ *is of type* S if $\mathcal{B} \in S(\Lambda)$ for some set Λ.

The operators ℓ^+ and ℓ^-, defined in Chapter 3 by (3.7) for a string are of type S. They belong to $S(\{\ell\})$, since they may be expressed as

$$\ell^\pm = \frac{\tau_\ell \pm \tau_{-\ell}}{2}.$$

We write these formulas in a unified way as

$$\ell^\varepsilon = \frac{\tau_\ell + \varepsilon \tau_{-\ell}}{2}, \tag{5.21}$$

where $\varepsilon = \pm 1$.

In the following proposition we gather two elementary properties of the operators of type S. The operators $\mathcal{P}$ and $\mathcal{Q}$, which we will construct for the network with the property (5.19), will be products of the operators $\ell_i^\pm$ constructed for the lengths ℓ_i of the strings involved in the network.

The following proposition shows why it is natural to consider the class of operators $S(\Lambda)$ to characterize the operators $\mathcal{P}$ and $\mathcal{Q}$.

Proposition 5.1. *(i)* $\mathcal{B} \in S(\Lambda)$ *if, and only if, it may be written as a linear combination of operators of the form* $\ell_1^{\varepsilon_1} \ell_2^{\varepsilon_2} \cdots \ell_n^{\varepsilon_n}$, *where each ε_i is -1 or 1.*

(ii) If $\mathcal{B}_1 \in S(\Lambda_1)$ and $\mathcal{B}_2 \in S(\Lambda_2)$ then, $\mathcal{B}_1 \mathcal{B}_2 = \mathcal{B}_2 \mathcal{B}_1 \in S(\Lambda_1 \sqcup \Lambda_2)$ and $s(\mathcal{B}_1 \mathcal{B}_2) = s(\mathcal{B}_1) + s(\mathcal{B}_2)$.

Proof. These properties are based on the fact that, if $\alpha, \beta \in \mathbb{R}$ then,

$$\tau_\alpha \tau_\beta = \tau_{\alpha+\beta} = \tau_\beta \tau_\alpha.$$

This implies, in particular, that the operators $\ell_i^{\varepsilon_i}$ and $\ell_j^{\varepsilon_j}$ commute.

In account of (5.21), a product of the form $\ell_1^{\varepsilon_1} \ell_2^{\varepsilon_2} \cdots \ell_n^{\varepsilon_n}$ may be expressed as

$$\ell_1^{\varepsilon_1} \ell_2^{\varepsilon_2} \cdots \ell_n^{\varepsilon_n} = \prod_{i=1}^{n} \left(\frac{\tau_{\ell_i} + \varepsilon_i \tau_{-\ell_i}}{2} \right);$$

then we get

$$\ell_1^{\varepsilon_1} \ell_2^{\varepsilon_2} \cdots \ell_n^{\varepsilon_n} = \sum_{h \in \mathcal{H}_\Lambda} c_h \tau_h \in S(\Lambda).$$

Thus, any linear combination of the operators $\ell_1^{\varepsilon_1} \ell_2^{\varepsilon_2} \cdots \ell_n^{\varepsilon_n}$ is an operator of type $S(\Lambda)$.

Conversely, if $h \in \mathcal{H}_\Lambda$, then

$$h = \sum_{i=1}^{n} \varepsilon_i \ell_i, \qquad \varepsilon_i = \pm 1,$$

and in view of the fact that, as it follows from (5.21), for any ℓ, $\tau_{\varepsilon \ell} = \varepsilon \ell^\varepsilon + \ell^{-\varepsilon}$,

$$\tau_h = \prod_{i=1}^{n} \tau_{\varepsilon_i \ell_i} = \prod_{i=1}^{n} \left[\varepsilon_i \ell_i^{\varepsilon_i} + \ell_i^{-\varepsilon_i} \right].$$

From here, τ_h may be expressed as a sum of products of the form $\ell_1^{\varepsilon_1} \ell_2^{\varepsilon_2} \cdots \ell_n^{\varepsilon_n}$ and the same will be true for each $\mathcal{B} = \sum_{h \in \mathcal{H}_\Lambda} c_h \tau_h$.

The assertion (ii) follows immediately. If

$$\Lambda_1 = \{\ell_1, \ldots, \ell_n\}$$

and

$$\Lambda_2 = \{\ell_{n+1}, \ldots, \ell_N\}$$

then,

$$\Lambda_1 \sqcup \Lambda_2 = \{\ell_1, \ldots, \ell_N\}.$$

If $\mathcal{B}_1 = \ell_1^{\varepsilon_1} \cdots \ell_n^{\varepsilon_n} \in S(\Lambda_1)$, $\mathcal{B}_2 = \ell_{n+1}^{\varepsilon_{n+1}} \cdots \ell_N^{\varepsilon_N} \in S(\Lambda_2)$,

$$\mathcal{B}_1 \mathcal{B}_2 = \mathcal{B}_2 \mathcal{B}_1 = \ell_1^{\varepsilon_1} \cdots \ell_n^{\varepsilon_n} \ell_{n+1}^{\varepsilon_1} \cdots \ell_N^{\varepsilon_N} \in S(\Lambda_1 \sqcup \Lambda_2).$$

Taking now into account that any $\mathcal{B}_1 \in S(\Lambda_1)$ and $\mathcal{B}_2 \in S(\Lambda_2)$ may be expressed, respectively, by means of linear combinations of the operators $\ell_1^{\varepsilon_1} \cdots \ell_n^{\varepsilon_n}$ and $\ell_{n+1}^{\varepsilon_{n+1}} \cdots \ell_N^{\varepsilon_N}$, (ii) is obtained in the general case.

In the rest of this chapter, when an operator $\mathcal{B}$ of type S is applied to a function w depending on the real variable t (and, possibly, on other variables), we will assume that $\mathcal{B}$ acts only on that variable t. In particular, if $w(t, x)$ is a function defined on $\mathbb{R} \times [a, b]$ then

$$\ell^{\pm} w(t, x) = \frac{1}{2} (w(t + \ell, x) \pm w(t - \ell, x)).$$

The following facts are widely used in the proof of our main results.

Proposition 5.2. *Let* $w(t,x)$ *be a function defined on* $\mathbb{R} \times [0, \ell]$. *Then,*

$$\mathbf{E}_{\ell^{\pm}w}(t) \leq \ell^{+} \mathbf{E}_w(t).$$

Proof.   For every $t \in \mathbb{R}$

$\mathbf{E}_{\ell^{\pm}w}(t)$

$$= \frac{1}{8} \int_0^{\ell} \left\{ |w_x(t+\ell,x) \pm w_x(t-\ell,x)|^2 + |w_t(t+\ell,x) \pm w_t(t-\ell,x)|^2 \right\} dx$$

$$\leq \frac{1}{4} \int_0^{\ell} \left\{ |w_x(t+\ell,x)|^2 + |w_x(t-\ell,x)|^2 + |w_t(t+\ell,x)|^2 + |w_t(t-\ell,x)|^2 \right\} dx$$

$$= \frac{1}{2} \left(\mathbf{E}_w(t+\ell) + \mathbf{E}_w(t-\ell) \right) = \ell^{+} \mathbf{E}_w(t).$$

Proposition 5.3. *If* $\mathcal{B}$ *is an operator of type* S *with* $s(\mathcal{B}) = s$ *then there exist positive constants* C_1, C_2, *depending only on the coefficients of* $\mathcal{B}$, *such that*
(i) inequality

$$\int_a^b |\mathcal{B}f(t)|^2 dt \leq C_1 \int_{a-s}^{b+s} |f(t)|^2 dt,$$

holds for all the functions f *for which both integrals are defined*[1].

   (ii) If the function $w(t,x)$ *is defined on* $\mathbb{R} \times [0, \ell]$ *and there exists a constant* $M > 0$ *such that* $\mathbf{E}_w(t) \leq M$ *for every* $t \in [a, b]$ *then* $\mathbf{E}_{\mathcal{B}w}(t) \leq C_2 M$ *for* $t \in [a+s, b-s]$.

Proof. (i) When the set Λ is formed by a single element: $\Lambda = \{\ell\}$, we have $\mathcal{B} = c_1 \ell^{+} + c_2 \ell^{-}$ and $s(\mathcal{B}) = \ell$. Then,

$$\int_a^b |\mathcal{B}f(t)|^2 dt = \int_a^b \left| c_1 \ell^{+} f(t) + c_2 \ell^{-} f(t) \right|^2 dt$$

$$= \int_a^b \left| \frac{c_1 + c_2}{2} f(t+\ell) + \frac{c_1 - c_2}{2} f(t-\ell) \right|^2 dt$$

$$\leq 2 \left(\frac{c_1 + c_2}{2} \right)^2 \int_a^b |f(t+\ell)|^2 dt + 2 \left(\frac{c_1 - c_2}{2} \right)^2 \int_a^b |f(t-\ell)|^2 dt$$

$$\leq 2 \left(\frac{c_1 + c_2}{2} \right)^2 \int_{a+\ell}^{b+\ell} |f(t)|^2 dt + 2 \left(\frac{c_1 - c_2}{2} \right)^2 \int_{a-\ell}^{b-\ell} |f(t)|^2 dt$$

$$\leq \left(c_1^2 + c_2^2 \right) \int_{a-\ell}^{b+\ell} |f(t)|^2 dt.$$

   When $n \geq 2$, it suffices to iterate this inequality taking under consideration Proposition 5.1(i). Let us note that C_1 may be chosen as the maximum of the squares of the coefficients of $\mathcal{B}$ in its representation given by Proposition 5.1(i) and then, C_1 depends only on $\mathcal{B}$.

   (ii) This is an immediate consequence of Proposition 5.2.

---

[1] In other words, $\mathcal{B}$ is continuous from $L^2[a-s, b+s]$ to $L^2[a,b]$.

The next proposition plays a crucial role in obtaining the optimal time in the observability inequalities that we prove for the solutions of the homogeneous system (5.1)-(5.6). Let us note that this fact was already proved in Section 4.4 of Chapter 4 for the operator $\mathcal{Q}$ corresponding to the three string network.

Proposition 5.4. *Let* $\Lambda = \{\ell_1, ..., \ell_m\}$ *with* $\ell_1 \leq ... \leq \ell_m$ *and denote* $T_\Lambda = 2s(\Lambda) = 2\sum_{i=1}^{m} \ell_i$. *Assume that* $\mathcal{B} = \sum_{h \in \mathcal{H}_\Lambda} c_h \tau_h \in S(\Lambda)$ *and that the coefficient* $c_{\ell_1 + \cdots + \ell_m}$ *is different from zero. Then, for any* $T > 0$ *there exists a constant* $C_T > 0$ *such that*

$$\int_0^T |u(t)|^2 dt \leq C_T \int_0^{T_\Lambda} |u(t)|^2 dt$$

for any continuous function u *satisfying* $\mathcal{B}u \equiv 0$.

Proof. We shall prove that, for any natural number n and any function u satisfying $\mathcal{B}u \equiv 0$, it holds that

$$\int_0^{T_\Lambda + 2n\ell_1} |u(t)|^2 dt \leq \gamma^n \int_0^{T_\Lambda} |u(t)|^2 dt, \tag{5.22}$$

where γ is a positive constant depending only on $\mathcal{B}$. Clearly, the assertion of the proposition immediately follows from inequality (5.22).

If $\mathcal{B}u \equiv 0$, i.e., $0 = \sum_{h \in \mathcal{H}_\Lambda} c_h \tau_h u(t) = \sum_{h \in \mathcal{H}_\Lambda} c_h u(t + h)$, then, replacing the variable t by $t - (\ell_1 + \cdots \ell_m)$ and taking into account that $c_{\ell_1 + \cdots \ell_m} \neq 0$ we get

$$u(t) = \sum_{h' \in \mathcal{H}_\Lambda^*} \delta_{h'} u(t - h'), \tag{5.23}$$

where $\mathcal{H}_\Lambda^* = \{h' = h - (\ell_1 + \cdots \ell_m) : h \in \mathcal{H}_\Lambda, h \neq (\ell_1 + \cdots \ell_m)\}$ and

$$\delta_{h'} = -\frac{c_{h' + (\ell_1 + \cdots \ell_m)}}{c_{\ell_1 + \cdots \ell_m}}.$$

From (5.23) and the Cauchy-Schwartz inequality it follows

$$|u(t)|^2 \leq \delta \sum_{h' \in \mathcal{H}_\Lambda^*} |u(t - h')|^2, \tag{5.24}$$

where $\delta = \sum_{h' \in \mathcal{H}_\Lambda^*} \delta_{h'}^2$.

Note that, for every $h' \in \mathcal{H}_\Lambda^*$ we have $2\ell_1 \leq h' \leq 2(\ell_2 + \cdots + \ell_m)$, and therefore,

$$T_\Lambda + 2(n+1)\ell_1 - h' \leq T_\Lambda + 2n\ell_1 \quad \text{and} \quad T_\Lambda + 2n\ell_1 - h' \geq 2(n+1)\ell_1 \geq 0.$$

This fact implies that

$$\int_{T_\Lambda+2n\ell_1-h'}^{T_\Lambda+2(n+1)\ell_1-h'} |u(t)|^2 dt \leq \int_0^{T_\Lambda+2n\ell_1} |u(t)|^2 dt. \tag{5.25}$$

On the other hand, from (5.24) it follows that

$$\int_{T_\Lambda+2n\ell_1}^{T_\Lambda+2(n+1)\ell_1} |u(t)|^2 dt \leq \delta \sum_{h'\in\mathcal{H}_\Lambda^*} \int_{T_\Lambda+2n\ell_1}^{T_\Lambda+2(n+1)\ell_1} |u(t-h')|^2 dt$$

$$= \delta \sum_{h'\in\mathcal{H}_\Lambda^*} \int_{T_\Lambda+2n\ell_1-h'}^{T_\Lambda+2(n+1)\ell_1-h'} |u(t)|^2 dt.$$

Now, taking (5.25) into account, the previous inequality becomes

$$\int_{T_\Lambda+2n\ell_1}^{T_\Lambda+2(n+1)\ell_1} |u(t)|^2 dt \leq (2^m-1)\delta \int_0^{T_\Lambda+2n\ell_1} |u(t)|^2 dt.$$

From this latter inequality we obtain

$$\int_0^{T_\Lambda+2(n+1)\ell_1} |u(t)|^2 dt = \int_0^{T_\Lambda+2n\ell_1} |u(t)|^2 dt + \int_{T_\Lambda+2n\ell_1}^{T_\Lambda+2(n+1)\ell_1} |u(t)|^2 dt \leq$$

$$\leq \int_0^{T_\Lambda+2n\ell_1} |u(t)|^2 dt + (2^m-1)\delta \int_0^{T_\Lambda+2n\ell_1} |u(t)|^2 dt$$

$$\leq (1+(2^m-1)\delta) \int_0^{T_\Lambda+2n\ell_1} |u(t)|^2 dt,$$

which proves inequality (5.22) with $\gamma = 1 + (2^m-1)\delta$.

Remark 5.5. If $\mathcal{B}$ is an operator of type S there exists a unique function $b(\lambda)$ such that $\mathcal{B}e^{i\lambda t} = b(\lambda)e^{i\lambda t}$. Indeed, it suffices to express $\mathcal{B}$ in the form

$$\mathcal{B} = \sum_{m\in\{0,1\}^n} d_m \ell_1^{\varepsilon_1} \cdots \ell_n^{\varepsilon_n}, \tag{5.26}$$

given by Proposition 5.1(i), to see that

$$\mathcal{B}e^{i\lambda t} = \sum d_m \ell_1^{\varepsilon_1} \cdots \ell_n^{\varepsilon_n} e^{i\lambda t}.$$

Taking into account that

$$\ell^+ e^{i\lambda t} = \cos\ell\lambda \, e^{i\lambda t}, \qquad \ell^- e^{i\lambda t} = i\sin\ell\lambda \, e^{i\lambda t},$$

it holds

$$\mathcal{B}e^{i\lambda t} = \sum d_m L_1^{\varepsilon_1}(\lambda) \cdots L_n^{\varepsilon_n}(\lambda) e^{i\lambda t},$$

where $L_i^{\varepsilon_i}(\lambda) = \cos\ell_i\lambda$ if $\varepsilon_i = 1$ and $L_i^{\varepsilon_i}(\lambda) = i\sin\ell_i\lambda$ if $\varepsilon_i = -1$. This means that the function $b(\lambda)$ may be constructed by replacing the operators ℓ_i^+ and ℓ_i^- in the decomposition (5.26) of $\mathcal{B}$ by $\cos\lambda t$ and $i\sin\lambda t$, respectively.

The uniqueness of $b(\lambda)$ is immediate: if $\mathcal{B}e^{i\lambda t} = b(\lambda)e^{i\lambda t} = c(\lambda)e^{i\lambda t}$, then $b(\lambda) = c(\lambda)$.

5.2.3 Construction of $\mathcal{P}$ and $\mathcal{Q}$ in the General Case

The construction of $\mathcal{P}$ and $\mathcal{Q}$ will be done by induction. We remind that such operators have already been constructed for a network consisting of a single string.

We shall denote by Λ_i the set of all the lengths of the strings of the sub-tree $\mathcal{A}_i$ and by $\Lambda_{\mathcal{A}}$ that of all the lengths of the tree $\mathcal{A}$. Suppose that for the sub-trees $\mathcal{A}_i$, $i = 1, \ldots, m$, we have already constructed the operators $\mathcal{P}_i$, $\mathcal{Q}_i$ that belong to $S(\Lambda_i)$ and verify

$$\mathcal{P}_i G_i + \mathcal{Q}_i F_i = 0, \tag{5.27}$$

where G_i and F_i are the velocity and the tension at the root of the sub-tree $\mathcal{A}_i$, i.e., at the vertex $\mathcal{O}$ of $\mathcal{A}$.

We define the operators

$$\mathcal{P} = \ell^+ \sum_{i=1}^{m} \mathcal{P}_i \prod_{j \neq i} \mathcal{Q}_j + \ell^- \prod_{j=1}^{m} \mathcal{Q}_j, \tag{5.28}$$

$$\mathcal{Q} = \ell^- \sum_{i=1}^{m} \mathcal{P}_i \prod_{j \neq i} \mathcal{Q}_j + \ell^+ \prod_{j=1}^{m} \mathcal{Q}_j. \tag{5.29}$$

Here and in the sequel the products denote the composition of operators.

Those are precisely the operators we are looking for.

Proposition 5.6. *The operators $\mathcal{P}$ and $\mathcal{Q}$ defined by (5.28)-(5.29) belong to $S(\Lambda_{\mathcal{A}})$. If $\bar{u}$ is a solution of (N) then*

$$\mathcal{P}G + \mathcal{Q}F = 0.$$

Proof. To prove that $\mathcal{P}, \mathcal{Q} \in S(\Lambda_{\mathcal{A}})$, it suffices to observe that, according to Proposition 5.1, all the terms of the sums in (5.28) and (5.29) belong to $S(\{\ell\} \sqcup \Lambda_1 \sqcup \ldots \sqcup \Lambda_m) = S(\Lambda_{\mathcal{A}})$. Using (5.17)-(5.18), the coupling conditions (5.16) between the strings may be expressed as

$$\sum_{i=1}^{m} F_i = \ell^- G + \ell^+ F, \qquad G_i = \ell^+ G + \ell^- F, \quad i = 1, \ldots, m. \tag{5.30}$$

From (5.28)-(5.29) we have

$$\mathcal{P}G + \mathcal{Q}F = \sum_{i=1}^{m} (\mathcal{P}_i \prod_{j \neq i} \mathcal{Q}_j) \ell^+ G + \prod_{j=1}^{m} \mathcal{Q}_j \ell^- G + \sum_{i=1}^{m} (\mathcal{P}_i \prod_{j \neq i} \mathcal{Q}_j) \ell^- F + \prod_{j=1}^{m} \mathcal{Q}_j \ell^+ F$$

$$= \sum_{i=1}^{m} (\mathcal{P}_i \prod_{j \neq i} \mathcal{Q}_j)(\ell^+ G + \ell^- F) + \prod_{j=1}^{m} \mathcal{Q}_j (\ell^- G + \ell^+ F).$$

Then, using formulas (5.30),

$$\mathcal{P}G + \mathcal{Q}F = \sum_{i=1}^{m}(\mathcal{P}_i \prod_{j \neq i} \mathcal{Q}_j)G_i + \sum_{i=1}^{m}(\prod_{j=1}^{m} \mathcal{Q}_j)F_i = \sum_{i=1}^{m}(\prod_{j \neq i} \mathcal{Q}_j)(\mathcal{P}_i G_i + \mathcal{Q}_i F_i) = 0,$$

where the last equality follows from the hypotheses (5.27). Thus, $\mathcal{P}$ and $\mathcal{Q}$, defined by (5.28)-(5.29), satisfy the relation (5.19).

Remark 5.7. From the definition, all $S(\Lambda)$-operator $\mathcal{B}$ may be written in the form

$$\mathcal{B} = \sum_{h \in \mathcal{H}_\Lambda} c_h \tau_h. \tag{5.31}$$

In general, this representation is not unique, since some elements of $\mathcal{H}_\Lambda$ may coincide. However, the coefficient $c_{s(\mathcal{B})} = c_{\ell_1 + \cdots + \ell_m}$, corresponding to the largest value of h, is determined in a unique way, as $\ell_1 + \cdots + \ell_m$ cannot coincide with another element of $\mathcal{H}_\Lambda$. Besides, it is easy to see that $c_{s(\mathcal{B})}$ is a multiplicative function, i.e., if $\mathcal{B}_1$ and $\mathcal{B}_2$ are S-operators with $s(\mathcal{B}_1) = s_1$ and $s(\mathcal{B}_2) = s_2$ then

$$c_{s_1+s_2}(\mathcal{B}_1\mathcal{B}_2) = c_{s_1}(\mathcal{B}_1)c_{s_2}(\mathcal{B}_2).$$

In the next proposition we study this coefficient for the operators $\mathcal{P}$ and $\mathcal{Q}$.

Proposition 5.8. *Let* $c_{L_\Lambda}(\mathcal{B})$ *denote the coefficient, corresponding to* $h = s(\Lambda_\Lambda) = L_\Lambda \in \mathcal{H}_{\Lambda_\Lambda}$ *in the expansion (5.31) of a* $S(\Lambda_\Lambda)$-*operator* $\mathcal{B}$. *Then* $c_{L_\Lambda}(\mathcal{P}) = c_{L_\Lambda}(\mathcal{Q}) > 0$.

Proof. We proceed by induction. For a string,

$$\mathcal{P} = \ell^+ = \frac{\tau_h + \tau_{-h}}{2}, \qquad \mathcal{Q} = \ell^+ = \frac{\tau_h - \tau_{-h}}{2}.$$

This implies $c_\ell(\mathcal{P}) = c_\ell(\mathcal{Q}) = 1/2$.

Now assume the assertion is true for the sub-trees $\mathcal{A}_1, ..., \mathcal{A}_m$. It means that

$$c_{L_i}(\mathcal{P}) = c_{L_i}(\mathcal{Q}) > 0, \quad i = 1, ..., m, \tag{5.32}$$

where, as above, L_i is the sum of the lengths of all the strings of the sub-tree $\mathcal{A}_i$.

Then, from formula (5.28) and the assumption (5.32)

$$c_{L_\Lambda}(\mathcal{P}) = c_{L_\Lambda}(\ell^+ \sum_{i=1}^{m} \mathcal{P}_i \prod_{j \neq i} \mathcal{Q}_j + \ell^- \prod_{j=1}^{m} \mathcal{Q}_j)$$

$$= c_{L_\Lambda}(\ell^+ \sum_{i=1}^{m} \mathcal{P}_i \prod_{j \neq i} \mathcal{Q}_j) + c_{L_\Lambda}(\ell^- \prod_{j=1}^{m} \mathcal{Q}_j)$$

$$= c_\ell(\ell^+) \sum_{i=1}^m c_{L_i}(\mathcal{P}_i) \prod_{j \neq i} c_{L_j}(\mathcal{Q}_j) + c_\ell(\ell^-) \prod_{j=1}^m c_{L_j}(\mathcal{Q}_j)$$

$$= \frac{1}{2}(m+1) \prod_{j=1}^m c_{L_j}(\mathcal{Q}_j) > 0.$$

In the same way it may be proved that

$$c_{L_A}(\mathcal{Q}) = \frac{1}{2}(m+1) \prod_{j=1}^m c_{L_j}(\mathcal{Q}_j),$$

what completes the proof.

5.2.4 The Action of $\mathcal{P}$ and $\mathcal{Q}$ at the Interior Nodes

For the index $\bar{\alpha} = (\alpha_1, ..., \alpha_k) \in \mathcal{J}$ we denote

$$\tilde{\Lambda}_{\bar{\alpha}} := \{\ell, \ell_{\alpha_1}, \ell_{\alpha_1, \alpha_2}, ..., \ell_{\alpha_1, \alpha_2, ..., \alpha_{k-1}}\}.$$

Observe that $\tilde{\Lambda}_{\bar{\alpha}}$ is the set of the lengths of the strings forming the unique simple path that connects the root $\mathcal{R}$ with the sub-tree $\mathcal{A}_{\bar{\alpha}}$. For completeness we take for the empty index $\tilde{\Lambda} = \emptyset$.

The following proposition gives information on how the operators $\mathcal{P}$ and $\mathcal{Q}$ act on traces of the components of a solution at the interior nodes of the network.

Proposition 5.9. *For any $\bar{\alpha} \in \mathcal{J}$ there exist operators $\mathcal{L}_{\bar{\alpha}} \in S(\Lambda_A \sqcup \tilde{\Lambda}_{\bar{\alpha}})$ such that, for any solution of* (N)

$$\mathcal{Q}F_{\bar{\alpha}} = \mathcal{L}_{\bar{\alpha}}G, \qquad \mathcal{P}F_{\bar{\alpha}} = -\mathcal{L}_{\bar{\alpha}}F.$$

Proof. We proceed by induction. Note that from the relation $\mathcal{P}G + \mathcal{Q}F = 0$, it follows that, when $\bar{\alpha}$ is the empty multi-index, the property is true with $\mathcal{L} = -\mathcal{P} \in S(\Lambda_A) = S(\Lambda_A \sqcup \tilde{\Lambda})$. In particular, for a single string the assertion of the proposition holds.

Suppose now that the operators $\mathcal{L}_{\bar{\alpha}}$ have been already constructed for the sub-trees $\mathcal{A}_i$, $i = 1, ..., m$, of $\mathcal{A}$. This means that for $i = 1, ..., m$, we have the operators $\mathcal{L}_{\bar{\alpha}}^i \in S(\Lambda_i \sqcup \tilde{\Lambda}_{\bar{\alpha}}^i)$ such that

$$\mathcal{P}_i F_{io\bar{\alpha}} = -\mathcal{L}_{\bar{\alpha}}^i F_i, \qquad \mathcal{Q}_i F_{io\bar{\alpha}} = \mathcal{L}_{\bar{\alpha}}^i G_i,$$

where $\tilde{\Lambda}_{\bar{\alpha}}^i$ is the set defined as $\tilde{\Lambda}_{\bar{\alpha}}$ for the sub-tree $\mathcal{A}_i$ and $\mathcal{P}_i$, $\mathcal{Q}_i$ are the operators $\mathcal{P}$, $\mathcal{Q}$ corresponding to that sub-tree.

Then, using relation (5.29),

$$\mathcal{Q}F_{io\bar{\alpha}} = \ell^- (\sum_{j=1}^m \mathcal{P}_j \prod_{k \neq j} \mathcal{Q}_k) F_{io\bar{\alpha}} + \ell^+ (\prod_{k=1}^m \mathcal{Q}_k) F_{io\bar{\alpha}}$$

$$= \ell^- (\sum_{\substack{j=1 \\ j \neq i}}^m \mathcal{P}_j \prod_{\substack{k \neq j \\ k \neq i}} \mathcal{Q}_k) \mathcal{Q}_i F_{io\bar{\alpha}} + \ell^- (\mathcal{P}_i \prod_{k \neq i} \mathcal{Q}_k) F_{io\bar{\alpha}} + \ell^+ (\prod_{k=1}^m \mathcal{Q}_k) F_{io\bar{\alpha}}$$

$$= \mathcal{L}_{\bar{\alpha}}^i \left(\ell^- (\sum_{\substack{i=1 \\ j \neq i}}^m \mathcal{P}_j \prod_{\substack{k \neq j \\ k \neq i}} \mathcal{Q}_k) G_i - \ell^- (\prod_{k \neq i} \mathcal{Q}_k) F_i + \ell^+ (\prod_{k=1}^m \mathcal{Q}_k) G_i \right)$$

$$= \mathcal{L}_{\bar{\alpha}}^i \left(\ell^- \sum_{j \neq i} (\prod_{\substack{k \neq j \\ k \neq i}} \mathcal{Q}_k) (\mathcal{P}_j G_i + \mathcal{Q}_j \widehat{F}) - \ell^- (\prod_{k \neq i} \mathcal{Q}_k) \widehat{F} + \ell^+ (\prod_{k \neq i} \mathcal{Q}_k) G_i \right)$$

$$= \mathcal{L}_{\bar{\alpha}}^i (\prod_{k \neq i} \mathcal{Q}_k) (\ell^+ \widehat{G} - \ell^- \widehat{F})$$

$$= \mathcal{L}_{\bar{\alpha}}^i (\prod_{k \neq i} \mathcal{Q}_k) \left(\ell^+ (\ell^- F + \ell^+ G) - \ell^- (\ell^+ F + \ell^- G) \right)$$

$$= \mathcal{L}_{\bar{\alpha}}^i (\prod_{k \neq i} \mathcal{Q}_k) \left((\ell^+)^2 - (\ell^-)^2 \right) G.$$

In a similar way, it may be obtained that

$$\mathcal{P}F_{io\bar{\alpha}} = - \mathcal{L}_{\bar{\alpha}}^i \prod_{k \neq i} \mathcal{Q}_k \left((\ell^+)^2 - (\ell^-)^2 \right) F.$$

Thus, we arrive to the recursive formula

$$\mathcal{L}_{io\bar{\alpha}} = \mathcal{L}_{\bar{\alpha}}^i \prod_{k \neq i} \mathcal{Q}_k \left((\ell^+)^2 - (\ell^-)^2 \right),$$

from which, in particular, according to Proposition 5.1, it holds that the operators $\mathcal{L}_{io\bar{\alpha}}$ belong to $S(\Lambda_i \sqcup \tilde{\Lambda}_{\bar{\alpha}}^i \sqcup \{\ell, \ell\}) = S(\Lambda_i \sqcup \{\ell\} \sqcup \tilde{\Lambda}_{\bar{\alpha}}^i \sqcup \{\ell\}) = S(\Lambda_A \sqcup \tilde{\Lambda}_{io\bar{\alpha}})$. This proves the proposition.

The action of $\mathcal{P}$ and $\mathcal{Q}$ on the velocities $G_{\bar{\alpha}}$ may be described in a similar way:

Proposition 5.10. *For any $\bar{\alpha} \in \mathcal{I}$ there exist operators $\mathcal{K}_{\bar{\alpha}}, \widehat{\mathcal{K}}_{\bar{\alpha}} \in S(\Lambda_A \sqcup \tilde{\Lambda}_{\bar{\alpha}})$ such that, for any solution of* (N)

$$\mathcal{Q}G_{\bar{\alpha}} = \mathcal{K}_{\bar{\alpha}}G, \qquad \mathcal{P}G_{\bar{\alpha}} = \widehat{\mathcal{K}}_{\bar{\alpha}}F.$$

Proof. From the relation $\mathcal{P}G + \mathcal{Q}F = 0$ it follows that for the empty multi-index $\mathcal{K} = \mathcal{Q}$ and $\widehat{\mathcal{K}} = -\mathcal{Q}$. For the remaining indices the operators $\mathcal{K}_{\bar{\alpha}}$ and $\widehat{\mathcal{K}}_{\bar{\alpha}}$ are constructed by recurrence. Assume that for the index $\bar{\alpha}$ the operators $\mathcal{K}_{\bar{\alpha}}$ and $\widehat{\mathcal{K}}_{\bar{\alpha}}$, verifying the conditions of the proposition, have been already constructed.

Then, for the indices $\bar{\alpha} \circ i$ with $i = 1, \ldots, m_{\bar{\alpha}}$, we have that

$$\mathcal{Q}G_{\bar{\alpha}\circ i} = \mathcal{Q}\widehat{G}_{\bar{\alpha}} = \ell_{\bar{\alpha}}^{+}\mathcal{Q}G_{\bar{\alpha}} + \ell_{\bar{\alpha}}^{-}\mathcal{Q}F_{\bar{\alpha}} = \left(\ell_{\bar{\alpha}}^{+}\mathcal{K}_{\bar{\alpha}} + \ell_{\bar{\alpha}}^{-}\mathcal{L}_{\bar{\alpha}}\right)G,$$

where $\mathcal{L}_{\bar{\alpha}}$ is the operator constructed in the previous proposition.

In an analogous way it may be obtained that

$$\mathcal{P}G_{\bar{\alpha}\circ i} = (\ell_{\bar{\alpha}}^{+}\widehat{\mathcal{K}}_{\bar{\alpha}} - \ell_{\bar{\alpha}}^{-}\mathcal{L}_{\bar{\alpha}})F.$$

Then, the needed operators may be constructed by the rules

$$\mathcal{K}_{\bar{\alpha}\circ i} = \ell_{\bar{\alpha}}^{+}\mathcal{K}_{\bar{\alpha}} + \ell_{\bar{\alpha}}^{-}\mathcal{L}_{\bar{\alpha}}, \tag{5.33}$$

$$\widehat{\mathcal{K}}_{\bar{\alpha}\circ i} = \ell_{\bar{\alpha}}^{+}\widehat{\mathcal{K}}_{\bar{\alpha}} - \ell_{\bar{\alpha}}^{-}\mathcal{L}_{\bar{\alpha}}. \tag{5.34}$$

As in the proof of Proposition 5.9, from the relations (5.33)-(5.34) it holds, in particular, that the operators $\mathcal{K}_{\bar{\alpha}\circ i}$ and $\widehat{\mathcal{K}}_{\bar{\alpha}\circ i}$ belong to $S(\Lambda_A \sqcup \tilde{\Lambda}_{\bar{\alpha}\circ i})$.

5.2.5 Action of $\mathcal{P}$ and $\mathcal{Q}$ on the Solution

If $\bar{\phi}$ is a solution of (N) and $\mathcal{B}$ is an operator of type S, then, due to the linearity of $\mathcal{B}$ and (N), $\mathcal{B}\bar{\phi}$ is also a solution of (N). Moreover, if $G_{\bar{\alpha}}^{\mathcal{B}\bar{\phi}}$ and $F_{\bar{\alpha}}^{\mathcal{B}\bar{\phi}}$, $\bar{\alpha} \in \mathcal{I}$, denote the traces of the velocity and tension of the strings at the vertices of the network for the solution $\mathcal{B}\bar{\phi}$, then

$$G_{\bar{\alpha}}^{\mathcal{B}\bar{\phi}} = \mathcal{B}G_{\bar{\alpha}}, \qquad F_{\bar{\alpha}}^{\mathcal{B}\bar{\phi}} = \mathcal{B}F_{\bar{\alpha}}.$$

That is true, in particular, when $\mathcal{B}$ is one of the operators $\mathcal{P}$ or $\mathcal{Q}$. The following lemma contains a fundamental technical step in our construction

Lemma 5.11. *There exists a constant C, independent of $\bar{\phi}$, such that*

$$\mathbf{E}_{\mathcal{P}\bar{\phi}}(t) \leq C \int_{T^*-2L_A}^{T^*+2L_A} |F(t)|^2 dt, \qquad \mathbf{E}_{\mathcal{Q}\bar{\phi}}(t) \leq C \int_{T^*-2L_A}^{T^*+2L_A} |G(t)|^2 dt \tag{5.35}$$

for every $T^* \in \mathbb{R}$ and $t \in [T^* - L_A, T^* + L_A]$.

Proof.   (i) Fix $T^* \in \mathbb{R}$. We shall prove first that

$$\mathbf{E}_{\mathcal{P}\bar{\phi}}(T^*) \leq C \int_{T^*-2L_A}^{T^*+2L_A} |F(t)|^2 dt, \; \mathbf{E}_{\mathcal{Q}\bar{\phi}}(T^*) \leq C \int_{T^*-2L_A}^{T^*+2L_A} |G(t)|^2 dt. \tag{5.36}$$

As a consequence of Propositions 5.9 and 5.10 we have

$$\mathcal{Q}F_{\bar{\alpha}} = \mathcal{L}_{\bar{\alpha}}G, \qquad \mathcal{Q}G_{\bar{\alpha}} = \mathcal{K}_{\bar{\alpha}}G,$$
$$\mathcal{P}F_{\bar{\alpha}} = -\mathcal{L}_{\bar{\alpha}}F, \qquad \mathcal{P}G_{\bar{\alpha}} = \widehat{\mathcal{K}}_{\bar{\alpha}}F,$$

for $\bar{\alpha} \in \mathcal{I}$. Then, from Propositions 3.1 and 5.3(i) it follows that

$$E_{\mathcal{Q}\bar{\phi}}^{\bar{\alpha}}(T^*) \le C \int_{T^*-\ell_{\bar{\alpha}}}^{T^*+\ell_{\bar{\alpha}}} \left(|\mathcal{L}_{\bar{\alpha}}G(t)|^2 + |K_{\bar{\alpha}}G(t)|^2\right) dt \le C \int_{T^*-2L_A}^{T^*+2L_A} |G(t)|^2 dt,$$

$$E_{\mathcal{P}\bar{\phi}}^{\bar{\alpha}}(T^*) \le C \int_{T^*-\ell_{\bar{\alpha}}}^{T^*+\ell_{\bar{\alpha}}} \left(|\mathcal{L}_{\bar{\alpha}}F(t)|^2 + |\widehat{\mathcal{K}}_{\bar{\alpha}}F(t)|^2\right) dt \le C \int_{T^*-2L_A}^{T^*+2L_A} |F(t)|^2 dt,$$

where, as above, $E^{\bar{\alpha}}$ is the energy of the solution in the string $\mathbf{e}_{\bar{\alpha}}$. It suffices to note that $\mathbf{E} = \sum_{\bar{\alpha}\in\mathcal{I}} E^{\bar{\alpha}}$ to obtain the inequalities (5.36).

(ii) Now we prove that they remain true for all $t \in [T^* - L_A, T^* + T_A]$. Indeed, if $t \in [T^* - L_A, T^* + L_A]$, from the formula (2.24) for the energy we have

$$\mathbf{E}_{\mathcal{P}\bar{\phi}}(t) = \mathbf{E}_{\mathcal{P}\bar{\phi}}(T^*) - \int_{T^*}^{t} F^{\mathcal{P}\bar{\phi}}(\tau)G^{\mathcal{P}\bar{\phi}}(\tau)dt$$

$$\le \mathbf{E}_{\mathcal{P}\bar{\phi}}(T^*) + \left|\int_{T^*}^{t} (|F^{\mathcal{P}\bar{\phi}}(\tau)|^2 + |G^{\mathcal{P}\bar{\phi}}(\tau)|^2)dt\right|$$

$$\le \mathbf{E}_{\mathcal{P}\bar{\phi}}(T^*) + \int_{T^*-L_A}^{T^*+L_A} (|\mathcal{P}F(\tau)|^2 + |\mathcal{P}G(\tau)|^2)dt$$

$$\le \mathbf{E}_{\mathcal{P}\bar{\phi}}(T^*) + \int_{T^*-L_A}^{T^*+L_A} (|\mathcal{P}F(\tau)|^2 + |\mathcal{Q}F(\tau)|^2)dt$$

$$\le C \int_{T^*-2L_A}^{T^*+2L_A} |F(\tau)|^2 dt$$

(in the last step we have used Proposition 5.3(i) and the result of the previous step (i) of this proof). For the operator $\mathcal{Q}$ the proof is similar.

Remark 5.12. When $\bar{\phi}$ is a solution of (5.7)-(5.11) (i.e., $G \equiv 0$), Lemma 5.11 gives $\mathbf{E}_{\mathcal{Q}\bar{\phi}}(t) = 0$. This implies that $\mathcal{Q}\bar{\phi}(t) = 0$. This relation may be viewed as a generalization of the time periodicity property of the solutions of the 1-d wave equation with homogeneous Dirichlet boundary conditions, which with our notations may be written as $\ell^- u(t) = 0$. As we have shown in Proposition 5.4, this generalized periodicity implies that all the essential L^2 information on $\bar{\phi}$ is contained in an interval of length $2L_A$.

5.3 The Main Observability Result

In this section we prove the main result on the observability of solutions of the homogeneous system (5.1)-(5.6).

For every non-empty multi-index $\bar{\alpha} = (\alpha_1, ..., \alpha_k) \in \mathcal{I}$ we define the operator $\mathcal{D}_{\bar{\alpha}}$ by

$$\mathcal{D}_{\bar{\alpha}} := \left(\prod_{i=1,\, i \neq \alpha_1}^{m} \mathcal{Q}_i \right) \left(\prod_{i=1,\, i \neq \alpha_2}^{m_{\alpha_1}} \mathcal{Q}_{\alpha_1, i} \right) \cdots \left(\prod_{i=1,\, i \neq \alpha_{k-1}}^{m_{\alpha_1, ..., \alpha_{k-1}}} \mathcal{Q}_{\alpha_1, ..., \alpha_{k-1}, i} \right) \tag{5.37}$$

with the convention that, for the empty index, $\mathcal{D}$ is the identity operator. We recall that $\mathcal{Q}_{\bar{\beta}}$ is the operator constructed in the previous section for the sub-tree $\mathcal{A}_{\bar{\beta}}$ and that the products in (5.37) denote the composition of operators.

Note that for every $\bar{\alpha} \in \mathcal{I}$ the operator $\mathcal{D}_{\bar{\alpha}}$ is of type S with $s(\mathcal{D}_{\bar{\alpha}}) < L_\mathcal{A}$.

The main observability result is as follows:

Theorem 5.13. *There exists a constant $C > 0$ such that*

$$\mathbf{E}_{\mathcal{D}_{\bar{\alpha}}\bar{\phi}}(0) = \mathbf{E}_{\mathcal{D}_{\bar{\alpha}}\bar{\phi}}(t) \leq C \int_0^{2L_\mathcal{A}} |F(\tau)|^2 d\tau,$$

for every solution $\bar{\phi}$ of (5.1)-(5.5) and every $\bar{\alpha} \in \mathcal{I}_S$.

The proof is based on

Lemma 5.14. *There exists a positive constant C, such that for every $\bar{\alpha} \in \mathcal{I}_S$ and every solution $\bar{\phi}$ of (N)*

$$\mathbf{E}_{\mathcal{D}_{\bar{\alpha}}\bar{\phi}}(t) \leq C \int_{t-2L_\mathcal{A}}^{t+2L_\mathcal{A}} \left(|F(\tau)|^2 + |G(\tau)|^2 \right) d\tau$$

for any $t \in \mathbb{R}$.

Proof. We proceed by induction. For the case of a single string the assertion is an immediate consequence of Proposition 3.1.

Now fix $\bar{\alpha} = (\alpha_1, ..., \alpha_k) \in \mathcal{I}_S$ and assume that the assertion of the theorem is true for the sub-tree $\mathcal{A}_{\alpha_1}$. That implies that

$$\mathbf{E}^{\alpha_1}_{\mathcal{D}^{\alpha_1}_{\bar{\alpha}}\bar{\phi}}(t) \leq C \int_{t-2L_{\alpha_1}}^{t+2L_{\alpha_1}} \left(|F_{\alpha_1}(\tau)|^2 + |G_{\alpha_1}(\tau)|^2 \right) d\tau \tag{5.38}$$

for any solution $\bar{\phi}$ of (N), where

$$\mathcal{D}^{\alpha_1}_{\bar{\alpha}} := \left(\prod_{i=1,\, i \neq \alpha_2}^{m_{\alpha_1}} \mathcal{Q}_{\alpha_1, i} \right) \cdots \left(\prod_{i=1,\, i \neq \alpha_{k-1}}^{m_{\alpha_1, ..., \alpha_{k-1}}} \mathcal{Q}_{\alpha_1, ..., \alpha_{k-1}, i} \right) \tag{5.39}$$

is the operator $\mathcal{D}_{\bar{\alpha}}$ for the sub-tree $\mathcal{A}_{\alpha_1}$ with $\bar{\alpha} = (\alpha_2, ..., \alpha_k)$.

First, we estimate the energy $\mathbf{E}^{\alpha_1}_{\mathcal{D}_{\bar{\alpha}}\bar{\phi}}$ of $\mathcal{D}_{\bar{\alpha}}\bar{\phi}$ on the sub-tree $\mathcal{A}_{\alpha_1}$. To do this, we set

$$\bar{\omega} := (\prod_{j=1,\, j\neq\alpha_1}^{m} \mathcal{Q}_j)\bar{\phi}, \qquad \bar{\omega}_i := (\prod_{\substack{j=1,\, j\neq\alpha_1 \\ j\neq i}}^{m} \mathcal{Q}_j)\bar{\phi}, \quad i = 1,\ldots,m. \qquad (5.40)$$

Note that these functions are also solutions of (N). They verify

$$\bar{\omega} = \mathcal{Q}_i\bar{\omega}_i, \qquad \mathcal{D}_{\bar{\alpha}}\bar{\phi} = \mathcal{D}_{\bar{\alpha}}^{\alpha_1}\bar{\omega}. \qquad (5.41)$$

Besides, from (5.38)

$$\mathbf{E}^{\alpha_1}_{\mathcal{D}_{\bar{\alpha}}^{\alpha_1}\bar{\omega}}(t) \leq C \int_{t-2L_{\alpha_1}}^{t+2L_{\alpha_1}} \left(|F^{\bar{\omega}}_{\alpha_1}(\tau)|^2 + |G^{\bar{\omega}}_{\alpha_1}(\tau)|^2 \right) d\tau. \qquad (5.42)$$

But, from the coupling formulas (5.16) we obtain that

$$\sum_{i=1}^{m} F_i^{\bar{\omega}} = \widehat{F}^{\bar{\omega}}, \qquad G_i^{\bar{\omega}} = \widehat{G}^{\bar{\omega}}, \qquad (5.43)$$

so that it holds

$$F^{\bar{\omega}}_{\alpha_1} = \widehat{F}^{\bar{\omega}} - \sum_{i=1,\, i\neq\alpha_1}^{m} \mathcal{Q}_i F_i^{\bar{\omega}_i} = \widehat{F}^{\bar{\omega}} + \sum_{i=1,\, i\neq\alpha_1}^{m} \mathcal{P}_i \widehat{G}^{\bar{\omega}_i}, \qquad G^{\bar{\omega}}_{\alpha_1} = \widehat{G}^{\bar{\omega}}. \quad (5.44)$$

Then, using the equalities (5.44), (5.42) we get

$$\mathbf{E}^{\alpha_1}_{\mathcal{D}_{\bar{\alpha}}^{\alpha_1}\bar{\omega}}(t) \leq C \int_{t-2L_{\alpha_1}}^{t+2L_{\alpha_1}} \left(|\widehat{F}^{\bar{\omega}}(\tau) + \sum_{i=1,\, i\neq\alpha_1}^{m} \mathcal{P}_i \widehat{G}^{\bar{\omega}_i}(\tau)|^2 + |\widehat{G}^{\bar{\omega}}(\tau)|^2 \right) d\tau$$

and this implies

$$\mathbf{E}^{\alpha_1}_{\mathcal{D}_{\bar{\alpha}}^{\alpha_1}\bar{\omega}}(t) \leq C \int_{t-2L_{\alpha_1}}^{t+2L_{\alpha_1}} \left(|\widehat{F}^{\bar{\omega}}(\tau)|^2 + \sum_{i=1,\, i\neq\alpha_1}^{m} |\mathcal{P}_i \widehat{G}^{\bar{\omega}_i}(\tau)|^2 + |\widehat{G}^{\bar{\omega}}(\tau)|^2 \right) d\tau.$$
$$(5.45)$$

Now, from the definition of $\bar{\omega}$ and the formulas (5.17), (5.18) we have

$$\widehat{F}^{\bar{\omega}} = (\prod_{j=1,\, j\neq\alpha_1}^{m} \mathcal{Q}_j)\widehat{F} = (\prod_{j=1,\, j\neq\alpha_1}^{m} \mathcal{Q}_j)\ell^+ F + (\prod_{j=1,\, j\neq\alpha_1}^{m} \mathcal{Q}_j)\ell^- G$$

and consequently

$$\int_{t-2L_{\alpha_1}}^{t+2L_{\alpha_1}} |\widehat{F}^{\bar{\omega}}|^2 d\tau$$

$$\leq 2 \int_{t-2L_{\alpha_1}}^{t+2L_{\alpha_1}} \left(|(\prod_{j=1,\, j\neq\alpha_1}^{m} \mathcal{Q}_j)\ell^+ F|^2 + |(\prod_{j=1,\, j\neq\alpha_1}^{m} \mathcal{Q}_j)\ell^- G|^2 \right) d\tau.$$

Observe that the operators $(\prod_{j=1,\, j\neq\alpha_1}^{m} \mathcal{Q}_j)\ell^+$ and $(\prod_{j=1,\, j\neq\alpha_1}^{m} \mathcal{Q}_j)\ell^-$ are of type S with $s < L_A - L_{\alpha_1}$ so that, the latter inequality combined with Proposition 5.3 provides

$$\int_{t-2L_{\alpha_1}}^{t+2L_{\alpha_1}} |\widehat{F}^{\bar{\omega}}(\tau)|^2 d\tau \leq C \int_{t-2L_A}^{t+2L_A} (|F(\tau)|^2 + |G(\tau)|^2) d\tau.$$

In a similar way it may be proved that

$$\int_{t-2L_{\alpha_1}}^{t+2L_{\alpha_1}} |\mathcal{P}_i \widehat{G}^{\bar{\omega}_i}(\tau)|^2 d\tau \leq C \int_{t-2L_A}^{t+2L_A} (|F(\tau)|^2 + |G(\tau)|^2) d\tau$$

and

$$\int_{t-2L_{\alpha_1}}^{t+2L_{\alpha_1}} |\widehat{G}^{\bar{\omega}}(\tau)|^2 d\tau \leq C \int_{t-2L_A}^{t+2L_A} (|F(\tau)|^2 + |G(\tau)|^2) d\tau.$$

Therefore, these three inequalities together with (5.41) and (5.45) give

$$\mathbf{E}_{\mathcal{D}_{\bar{\alpha}}\bar{\phi}}^{\alpha_1}(t) \leq C \int_{t-2L_A}^{t+2L_A} (|F(\tau)|^2 + |G(\tau)|^2) d\tau. \tag{5.46}$$

Now we proceed to estimate the energies $\mathbf{E}_{\mathcal{D}_{\bar{\alpha}}\bar{\phi}}^{i}$ of $\mathcal{D}_{\bar{\alpha}}\bar{\phi}$ on the remaining sub-trees $\mathcal{A}_i$ (i.e., for $i \neq \alpha_1$). According to Lemma 5.11, applied to $\bar{\omega}_i$ in the sub-tree $\mathcal{A}_i$, it holds that for every t' in $[t - L_i, t + L_i]$

$$\mathbf{E}_{\bar{\omega}}^{i}(t') = \mathbf{E}_{\mathcal{Q}_i \bar{\omega}_i}^{i}(t') \leq C \int_{t-2L_i}^{t+2L_i} |G_i^{\bar{\omega}_i}(\tau)|^2 d\tau, \tag{5.47}$$

for $i = 1, \ldots, m$. Taking into account that

$$G_i^{\bar{\omega}_i} = (\prod_{\substack{j=1,\, j\neq\alpha_1 \\ j\neq i}}^{m} \mathcal{Q}_j)\widehat{G} = (\prod_{\substack{j=1,\, j\neq\alpha_1 \\ j\neq i}}^{m} \mathcal{Q}_j)\ell^+ F + (\prod_{\substack{j=1,\, j\neq\alpha_1 \\ j\neq i}}^{m} \mathcal{Q}_j)\ell^- G, \tag{5.48}$$

we get from (5.47) and Proposition 5.3(i)

$$\mathbf{E}_{\bar{\omega}}^{i}(t') \leq C \int_{t-L_A-L_i+L_{\alpha_1}}^{t+L_A+L_i-L_{\alpha_1}} (|F(\tau)|^2 + |G(\tau)|^2) d\tau$$

$$\leq C \int_{t-2L_A}^{t+2L_A} (|F(\tau)|^2 + |G(\tau)|^2) d\tau. \tag{5.49}$$

Here we have used the fact that the operators applied to F and G in the right hand term of (5.48) are of type S with $s = L_A - L_{\alpha_1} - L_i$.

Now, if we apply Proposition 5.3(ii) with $\mathcal{B} = \mathcal{D}_{\bar{\alpha}}^{\alpha_1}$ to (5.49) (recall that $s(\mathcal{D}_{\bar{\alpha}}^{\alpha_1}) < L_{\alpha_1}$) we obtain, after choosing $t' = t$,

$$\mathbf{E}_{\mathcal{D}_{\bar{\alpha}}\bar{\phi}}^{i}(t) = \mathbf{E}_{\mathcal{D}_{\bar{\alpha}}^{\alpha_1}\bar{\omega}}^{i}(t') \leq C \int_{t-2L_A}^{t+2L_A} (|F(\tau)|^2 + |G(\tau)|^2) d\tau. \tag{5.50}$$

Finally, from Proposition 3.1 we obtain that the component ϕ of $\bar{\phi}$ verifies, for every $t' \in [t - L_{\mathcal{A}}, t + L_{\mathcal{A}}]$,

$$E_\phi(t') \leq C \int_{t-\ell-L_{\mathcal{A}}}^{t+\ell+L_{\mathcal{A}}} (|F(\tau)|^2 + |G(\tau)|^2)d\tau.$$

Thus, using Proposition 5.3(ii), it holds that

$$E_{\mathcal{D}_{\bar{\alpha}}\bar{\phi}}(t') \leq C \int_{t-\ell-L_{\mathcal{A}}}^{t+\ell+L_{\mathcal{A}}} (|F(\tau)|^2 + |G(\tau)|^2)d\tau,$$

for every $t' \in [t - L_{\mathcal{A}} + s(\mathcal{D}_{\bar{\alpha}}), t + L_{\mathcal{A}} - s(\mathcal{D}_{\bar{\alpha}})]$ and, since $s(\mathcal{D}_{\bar{\alpha}}) < L_{\mathcal{A}}$, this is true in particular for $t' = t$. Therefore,

$$E_{\mathcal{D}_{\bar{\alpha}}\bar{\phi}}(t) \leq C \int_{t-2L_{\mathcal{A}}}^{t+2L_{\mathcal{A}}} (|F(\tau)|^2 + |G(\tau)|^2)d\tau. \tag{5.51}$$

Now, it suffices to combine (5.46), (5.50), (5.51) and the fact that

$$\mathbf{E}_{\mathcal{D}_{\bar{\alpha}}\bar{\phi}} = E_{\mathcal{D}_{\bar{\alpha}}\bar{\phi}} + \sum_{i=1}^{m} \overset{i}{\mathcal{D}_{\bar{\alpha}}\bar{\phi}}$$

to conclude the proof.

With the help of Lemma 5.14 the proof of Theorem 5.13 is simple.

Proof (Proof of Theorem 5.13). If $\bar{\phi}$ is a solution of (5.1)-(5.6), so is $\mathcal{D}_{\bar{\alpha}}\bar{\phi}$. In particular, the energy of $\mathcal{D}_{\bar{\alpha}}\bar{\phi}$ is conserved. Then, taking into account that $G \equiv 0$ for the solutions of (5.1)-(5.5), from Lemma 5.14 it holds

$$\mathbf{E}_{\mathcal{D}_{\bar{\alpha}}\bar{\phi}}(0) = \mathbf{E}_{\mathcal{D}_{\bar{\alpha}}\bar{\phi}}(2T_{\mathcal{A}}) \leq C \int_0^{4L_{\mathcal{A}}} |F(\tau)|^2 d\tau. \tag{5.52}$$

On the other hand, in this case $\mathcal{Q}F \equiv 0$ and then, using Proposition 5.4 (which may be applied to $\mathcal{Q}$ on the basis of Proposition 5.8) we have

$$\int_0^{4L_{\mathcal{A}}} |F(\tau)|^2 d\tau \leq C \int_0^{2L_{\mathcal{A}}} |F(\tau)|^2 d\tau.$$

With this, the assertion of the theorem follows from (5.52).

5.4 Relation Between $\mathcal{P}$ and $\mathcal{Q}$ and the Spectrum

Our next objective is to express the inequality (5.52) in terms of the Fourier coefficients of the solution $\bar{\phi}$ of (5.7)-(5.12). This will lead to weighted observability inequalities with weights that depend on the eigenvalues μ_n of the operator $-\Delta_{\mathcal{A}}$. To study those weights we need some additional properties of the eigenvalues.

5.4.1 The Eigenvalue Problem

We consider the eigenvalue problem for the elliptic operator $-\Delta_A$ associated to the hyperbolic problem (5.7)-(5.11):

$$-\theta_{xx}^{\bar{\alpha}}(x) = \mu \, \theta^{\bar{\alpha}}(x) \qquad\qquad x \in [0, \ell_{\bar{\alpha}}], \quad \bar{\alpha} \in \mathfrak{I}, \tag{5.53}$$

$$\theta^{\bar{\alpha} \circ \beta}(0) = \theta^{\bar{\alpha}}(\ell_{\bar{\alpha}}) \qquad\qquad \bar{\alpha} \in \mathfrak{I}_{\mathcal{M}}, \quad \beta = 1, \ \ldots, m_{\bar{\alpha}}, \tag{5.54}$$

$$\sum_{\beta=1}^{m_{\bar{\alpha}}} \theta_x^{\bar{\alpha} \circ \beta}(0) = \theta_x^{\bar{\alpha}}(\ell_{\bar{\alpha}}) \qquad\qquad \bar{\alpha} \in \mathfrak{I}_{\mathcal{M}}, \tag{5.55}$$

$$\theta^{\bar{\alpha}}(\ell_{\bar{\alpha}}) = 0 \qquad\qquad \bar{\alpha} \in \mathfrak{I}_S, \tag{5.56}$$

$$\theta(0) = 0 \qquad\qquad \text{at the root } \mathcal{R}. \tag{5.57}$$

As it has been pointed out in Chapter 2, the spectrum of $-\Delta_A$ is formed by a positive, increasing sequence $\{\mu_k\}_{k \in \mathbb{Z}_+}$ of eigenvalues. We call it *spectrum of A* and denote it by σ_A.

Clearly, we may consider the problem (5.53)-(5.57) for each sub-tree $A_{\bar{\alpha}}$ of A. The corresponding spectrum is called *spectrum of $A_{\bar{\alpha}}$* and is denoted by $\sigma_{\bar{\alpha}}$.

For technical reasons, as we did for system (5.7)-(5.11), we will also consider smooth solutions of (5.53), which verify the boundary conditions (5.54)-(5.56) but not necessarily (5.57). For brevity, they are simply called *solutions of (N_E) corresponding to μ*.

Proposition 5.15. *If μ is a common eigenvalue of two sub-trees $A_{\bar{\alpha} \circ i}$, $A_{\bar{\alpha} \circ j}$ ($i \neq j$) with the same root $\mathbb{O}_{\bar{\alpha}}$ then μ is also an eigenvalue of A. Moreover, there exists a non-zero eigenfunction $\bar{\theta}$ associated to μ such that*

$$\theta(0) = \theta_x(0) = 0.$$

Proof. Let $\bar{\theta}^{\bar{\alpha} \circ i}$, $\bar{\theta}^{\bar{\alpha} \circ j}$ be non-zero eigenfunctions corresponding to the eigenvalue μ for the sub-trees $A_{\bar{\alpha} \circ i}$ and $A_{\bar{\alpha} \circ j}$, respectively. These functions are defined in the corresponding sub-trees but it will be sufficient to paste them conveniently to build up an eigenfunction of A.

We may assume that the numbers $\theta_x^{\bar{\alpha} \circ i}(0)$, $\theta_x^{\bar{\alpha} \circ j}(0)$ are both different from zero. Indeed, if one of them, say $\theta_x^{\bar{\alpha} \circ i}(0)$, vanishes then the relations

$$\theta^{\bar{\alpha} \circ i}(0) = \theta_x^{\bar{\alpha} \circ i}(0) = 0$$

allow to ensure that the function $\bar{\theta}$, obtained by extending by zero the function $\bar{\theta}^{\bar{\alpha} \circ i}$ to the whole tree A, satisfies (5.53)-(5.57) for that value of μ and then is an eigenfunction of A.

Now define the function $\bar{\theta}$ by

$$\theta_{\bar{\alpha}'} = \begin{cases} \theta_x^{\bar{\alpha} \circ j}(0) \, \theta_{\bar{\beta}}^{\bar{\alpha} \circ i} & \text{if } \bar{\alpha}' = \bar{\alpha} \circ i \circ \bar{\beta}, \\ -\theta_x^{\bar{\alpha} \circ i}(0) \, \theta_{\bar{\beta}}^{\bar{\alpha} \circ j} & \text{if } \bar{\alpha}' = \bar{\alpha} \circ j \circ \bar{\beta}, \\ 0 & \text{otherwise,} \end{cases}$$

i.e., $\bar{\theta}$ coincides in the sub-tree $\mathcal{A}_{\bar{\alpha}oi}$ with $\theta_x^{\bar{\alpha}oj}(0)\bar{\theta}^{\bar{\alpha}oi}$, in $\mathcal{A}_{\bar{\alpha}oj}$ with $-\theta_x^{\bar{\alpha}oi}(0)\bar{\theta}^{\bar{\alpha}oj}$ and vanishes outside those sub-trees. It is easy to see that $\bar{\theta}$ satisfies the boundary conditions (5.54)-(5.55) at $\mathcal{O}_{\bar{\alpha}}$:

$$\sum_{k=1}^{m_{\bar{\alpha}}} \theta_x^{\bar{\alpha}ok}(0) = \theta_x^{\bar{\alpha}oj}(0)\theta_x^{\bar{\alpha}oi}(0) - \theta_x^{\bar{\alpha}oi}(0)\theta_x^{\bar{\alpha}oj}(0) = 0 = \theta_x^{\bar{\alpha}}(\ell_{\bar{\alpha}}).$$

As at the other nodes they are obviously satisfied, $\bar{\theta}$ is an eigenfunction of $\mathcal{A}$.

Finally observe that in both cases, the eigenfunction $\bar{\theta}$ constructed here is such that

$$\theta(0) = \theta_x(0) = 0,$$

and thus, $\theta \equiv 0$, i.e., $\bar{\theta}$ vanishes at the whole string containing the root of $\mathcal{A}$.

Remark 5.16. Note that the eigenfunction constructed in the proof of Proposition 5.15 vanishes everywhere outside the sub-trees $\mathcal{A}_{\bar{\alpha}oi}$, $\mathcal{A}_{\bar{\alpha}oj}$. If we denote $\mathcal{A}_{\bar{\alpha}oi} \vee \mathcal{A}_{\bar{\alpha}oj}$ the tree formed by $\mathcal{A}_{\bar{\alpha}oi}$ and $\mathcal{A}_{\bar{\alpha}oj}$ in which the node $\mathcal{O}_{\bar{\alpha}}$ is considered as an interior point of a string of length $\ell_{\bar{\alpha}oi} + \ell_{\bar{\alpha}oj}$, we obtain that these sub-trees have a common eigenvalue if and only if there exists an eigenfunction of $\mathcal{A}_{\bar{\alpha}oi} \vee \mathcal{A}_{\bar{\alpha}oj}$ that vanishes at the point $\mathcal{O}_{\bar{\alpha}}$.

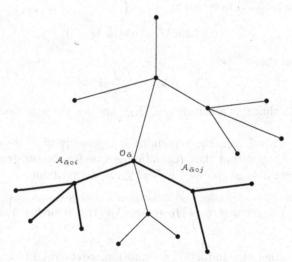

Fig. 5.2. The sub-tree $\mathcal{A}_{\bar{\alpha}oi} \vee \mathcal{A}_{\bar{\alpha}oj}$

As it has been shown above, the operators $\mathcal{P}$ and $\mathcal{Q}$ are of type S with $s(\mathcal{P}) = s(\mathcal{Q}) = L_{\mathcal{A}}$. According to Remark 5.5, there exist functions p and q such that for all $t, \lambda \in \mathbb{R}$,

$$\mathcal{P}e^{i\lambda t} = p(\lambda)e^{i\lambda t}, \qquad \mathcal{Q}e^{i\lambda t} = q(\lambda)e^{i\lambda t}. \tag{5.58}$$

Proposition 5.17. *Let $\lambda \in \mathbb{R} \setminus \{0\}$ and $f, g \in \mathbb{C}$ be such that*

$$q(\lambda)f + i\lambda p(\lambda)g = 0. \tag{5.59}$$

If the tree $\mathcal{A}$ satisfies the property

$$|q_{\bar{\alpha} \circ i}(\lambda)| + |q_{\bar{\alpha} \circ j}(\lambda)| \neq 0 \text{ for any } \bar{\alpha} \in \mathfrak{I}_{\mathcal{M}}, \ i, j = 1, ... m_{\bar{\alpha}}, \ i \neq j, \tag{5.60}$$

then there exists a unique solution $\bar{\theta}$ of (N_E) corresponding to the value $\mu = \lambda^2$ such that

$$\theta(0) = g \quad and \quad \theta_x(0) = f. \tag{5.61}$$

Proof. First we construct the component θ of $\bar{\theta}$ (the one corresponding to the string **e**). We set

$$\theta(x) = g \cos \lambda x + \frac{f}{\lambda} \sin \lambda x, \tag{5.62}$$

which clearly satisfies (5.61).

If the network consists of a single string of length ℓ, then

$$p(\lambda) = \cos \lambda \ell, \qquad q(\lambda) = i \sin \lambda \ell$$

and condition (5.59) becomes

$$if \sin \lambda \ell + ig\lambda \cos \lambda \ell = 0.$$

This implies that

$$\theta(\ell) = g \cos \lambda \ell + \frac{f}{\lambda} \sin \lambda \ell = 0,$$

what means that θ is a solution of (N_E) and so, the assertion is true in this case.

In the general case the remaining components of $\bar{\theta}$ are constructed by induction. Assume that the proposition is true for the sub-trees $\mathcal{A}_1, ..., \mathcal{A}_m$.

If we were able to choose numbers $f_1, ..., f_m$ verifying

$$\sum_{k=1}^{m} f_k = \theta_x(\ell) \text{ and } q_k(\lambda)f_k + i\lambda p_k(\lambda)\theta(\ell) = 0 \text{ for } k = 1, ..., m, \tag{5.63}$$

then, according to the induction assumption, we could find solutions $\bar{\theta}^1, ..., \bar{\theta}^m$, defined on the sub-trees $\mathcal{A}_1, ..., \mathcal{A}_m$, respectively, such that

$$\theta^k(0) = \theta(\ell), \quad \theta_x^k(0) = f_k, \text{ for } k = 1, ..., m.$$

This would imply that

$$\sum_{k=1}^{m} \theta_x^k(0) = \theta(\ell) \text{ and } \theta^k(0) = \theta(\ell), \text{ for } k = 1, ..., m.$$

Therefore, the function $\bar{\theta}$ defined on the tree by $\theta_{k \circ \bar{\alpha}} = \theta_{\bar{\alpha}}^k$ would be the solution of (N_E), whose existence is asserted in the proposition. Consequently, it remains to prove the possibility of making the decomposition (5.63).

We remark that from the definition of p and q and formulas (5.28), (5.29) it follows that

$$p = \cos \lambda \ell \sum_{k=1}^{m} p_k \prod_{j \neq k} q_j + i \sin \lambda \ell \prod_{j=1}^{m} q_j, \tag{5.64}$$

$$q = i \sin \lambda \ell \sum_{k=1}^{m} p_k \prod_{j \neq k} q_j + \cos \lambda \ell \prod_{j=1}^{m} q_j. \tag{5.65}$$

Note that condition (5.60) implies that among the numbers $q_k(\lambda)$, $k = 1, \dots, m$, at most one may be equal to zero. Thus, we consider two cases: a) all the numbers $q_k(\lambda)$, $k = 1, \dots, m$, are different from zero and b) exactly one of those numbers, say, e.g., $q_1(\lambda)$, is equal to zero.

Case a). If we take

$$f_k = \frac{-i \lambda p_k(\lambda) \theta(\ell)}{q_k(\lambda)},$$

then

$$\sum_{k=1}^{m} f_k = -i \lambda \theta(\ell) \sum_{k=1}^{m} \frac{p_k}{q_k} = -i \lambda (g \cos \lambda \ell + \frac{f}{\lambda} \sin \lambda \ell) \frac{\sum_{k=1}^{m} p_k \prod_{j \neq k} q_j}{\prod_{j=1}^{m} q_j}.$$

This equality, taking into account (5.64), (5.65), gives

$$\sum_{k=1}^{m} f_k = -i \lambda g \left(\frac{p}{\prod_{j=1}^{m} q_j} - i \sin \lambda \ell \right) - f \left(\frac{q}{\prod_{j=1}^{m} q_j} - \cos \lambda \ell \right)$$

$$= -\lambda g \sin \lambda \ell + f \cos \lambda \ell - \frac{i \lambda p g + q f}{\prod_{j=1}^{m} q_j} = -\lambda g \sin \lambda \ell + f \cos \lambda \ell = \theta_x(\ell).$$

Thus, the numbers $f_1, \dots, f_m$ satisfy (5.63).

Case b). The relations (5.64), (5.65) together with $q_1(\lambda) = 0$ give

$$p(\lambda) = \cos \lambda \ell p_1(\lambda) \prod_{j \neq 1} q_j(\lambda), \qquad q(\lambda) = i \sin \lambda \ell p_1(\lambda) \prod_{j \neq 1} q_j(\lambda),$$

and from (5.59) we obtain

$$0 = q(\lambda) f + i \lambda p(\lambda) g = i \lambda (g \cos \lambda \ell + \frac{g}{\lambda} \sin \lambda \ell) p_1(\lambda) \prod_{j \neq 1} q_j(\lambda)$$

$$= i \lambda \theta(\ell) p_1(\lambda) \prod_{j \neq 1} q_j(\lambda). \tag{5.66}$$

But $\prod_{j \neq 1} q_j(\lambda) \neq 0$ and then, necessarily, $\theta(\ell) p_1(\lambda) = 0$. It means that if we choose $f_1 = \theta_x(\ell)$ and $f_2, \dots, f_m$ verifying

$$\sum_{k=2}^{m} f_k = 0 \text{ and } q_k(\lambda)f_k + i\lambda p_k(\lambda)\theta(\ell) = 0 \text{ for } k = 2, ..., m,$$

as in the previous case then the condition (5.63) is satisfied.

So far, we have proved the existence of a solution. It turns out that for the solutions satisfying (5.17) we can give an explicit formula. Indeed, if we apply Propositions 5.9 and 5.10 to the solution

$$\bar{\theta}(t,x) = e^{i\lambda t}\bar{\theta}(x)$$

of (N) we obtain

$$\begin{aligned} i\lambda p(\lambda)\theta^{\bar{\alpha}}(0) &= \hat{k}(\lambda)\theta_x(0) = \hat{k}(\lambda)f, \\ p(\lambda)\theta_x^{\bar{\alpha}}(0) &= -l(\lambda)\theta_x(0) = -l(\lambda)f, \end{aligned} \qquad (5.67)$$

$$q(\lambda)\theta^{\bar{\alpha}}(0) = k(\lambda)\theta(0) = k(\lambda)g, \quad q(\lambda)\theta_x^{\bar{\alpha}}(0) = i\lambda l(\lambda)\theta(0) = i\lambda l(\lambda)g, \quad (5.68)$$

where k, $\hat{k}$, l and r are the functions associated to the operators $\mathcal{K}$, $\hat{\mathcal{K}}$, $\mathcal{L}$ and $\mathcal{R}$, respectively, according to Remark 5.5.

On the other hand, the condition (5.60) implies that at least one of the numbers $p(\lambda)$ or $q(\lambda)$ is different from zero (see Proposition 5.20 below). Therefore, one of the equalities (5.67), (5.68) provides us with an explicit formula for the values of $\theta^{\bar{\alpha}}(0)$ and $\theta_x^{\bar{\alpha}}(0)$ for any $\bar{\alpha} \in \mathfrak{I}_M$ and thus, for the solution $\bar{\theta}$. In particular, if $f = g = 0$ the corresponding solution vanishes identically on $\mathcal{A}$, what clearly implies the uniqueness of the solution for arbitrary values of f and g.

Remark 5.18. The converse assertion is also true, even if the condition (5.60) is not fulfilled. Indeed, if $\bar{\theta}$ is a solution of (N_E) then

$$\bar{\theta}(t,x) = e^{i\lambda t}\bar{\theta}(x)$$

is a solution of (N) and

$$\theta_t(t,0) = i\lambda e^{i\lambda t}\theta(0), \quad \theta_x(t,0) = e^{i\lambda t}\theta_x(0).$$

Then, from the relations (5.19) and (5.58) it follows

$$0 = \mathcal{P}\theta_t(t,0) + \mathcal{Q}\theta_x(t,0) = (ip\lambda\theta(0) + q\theta_x(0))e^{i\lambda t},$$

for every $t \in \mathbb{R}$. Thus, (5.59) holds.

Now we are ready to prove the following basic property.

Proposition 5.19. *Let $0 \neq \lambda \in \mathbb{R}$. Then λ^2 is an eigenvalue of $\mathcal{A}$ if and only if $q(\lambda) = 0$.*

Proof. First we prove that $q(\lambda) = 0$ implies that λ^2 is an eigenvalue, i.e., that there exists a non-zero solution of (5.53)-(5.57) for that value of λ. If the tree verifies (5.60) then this fact follows immediately from Proposition 5.17 choosing $g = 0$, $f \neq 0$. Note that the condition $0 \neq f = \theta_x(0)$ guarantees that $\bar{\theta}$ is not identically equal to zero. In particular, the assertion is true for a string, as it always verifies (5.60).

In the general case when the condition (5.60) may fail, we follow an induction argument: we suppose that the assertion has been proved for all the sub-trees $\mathcal{A}_{\bar{\alpha}}$ with non-empty $\bar{\alpha}$.

If $q_{\bar{\alpha}\circ i}(\lambda) = q_{\bar{\alpha}\circ j}(\lambda) = 0$ for some $\bar{\alpha} \in \mathfrak{I}_{\mathcal{M}}$, $i \neq j$, then, according to the induction hypothesis, λ^2 is an eigenvalue of both $\mathcal{A}_{\bar{\alpha}\circ i}$ and $\mathcal{A}_{\bar{\alpha}\circ j}$. Then from Proposition 5.15 it follows that λ^2 is an eigenvalue of $\mathcal{A}$, too.

Let us see now the converse assertion. Let $\bar{\theta}$ be a non-zero eigenfunction corresponding to the eigenvalue λ^2. Then the function $\bar{u}(t, x) = e^{i\lambda t}\bar{\theta}(x)$ is a solution of (N). Choose $\bar{\alpha} \in \mathfrak{I}$ such that one of the numbers $\theta^{\bar{\alpha}}(0)$ or $\theta_x^{\bar{\alpha}}(0)$ is different from zero (that is possible since, otherwise, it would be $\bar{\theta} \equiv 0$). For this solution of (N) we have for every $\bar{\alpha} \in \mathfrak{I}$

$$F_{\bar{\alpha}}(t) = e^{i\lambda t}\theta_x^{\bar{\alpha}}(0), \qquad G_{\bar{\alpha}}(t) = i\lambda e^{i\lambda t}\theta^{\bar{\alpha}}(0),$$

and in particular, $G \equiv 0$. Then, from Propositions 5.9 and 5.10 it follows that

$$0' = \mathcal{L}_{\bar{\alpha}}G = \mathcal{Q}F_{\bar{\alpha}} = \mathcal{Q}e^{i\lambda t}\theta_x^{\bar{\alpha}}(0) = q(\lambda)\theta_x^{\bar{\alpha}}(0),$$

$$0 = \mathcal{K}_{\bar{\alpha}}G = \mathcal{Q}G_{\bar{\alpha}} = \mathcal{Q}e^{i\lambda t}\theta^{\bar{\alpha}}(0) = i\lambda q(\lambda)\theta^{\bar{\alpha}}(0),$$

and therefore, necessarily, $q(\lambda) = 0$.

5.4.2 Further Properties of $\mathcal{P}$ and $\mathcal{Q}$

Proposition 5.20. *For every tree $\mathcal{A}$ the following properties hold:*
(i) one of the functions p, q is even and the other one is odd;
(ii) there exists $\lambda_0 \in \mathbb{R}$ such that $p(\lambda_0) = q(\lambda_0) = 0$ if, and only if, there exist two sub-trees $\mathcal{A}_{\bar{\alpha}\circ i}$, $\mathcal{A}_{\bar{\alpha}\circ j}$, $i \neq j$, with common root $\mathcal{O}_{\bar{\alpha}}$ such that

$$q_{\bar{\alpha}\circ i}(\lambda_0) = q_{\bar{\alpha}\circ j}(\lambda_0) = 0.$$

Proof. We proceed by induction. For a single string

$$p(\lambda) = \cos \lambda\ell, \qquad q(\lambda) = i \sin \lambda\ell.$$

In this case (i) is trivial. Assertion (ii) follows from the fact that $|p|^2 + |q|^2 = 1$.

Suppose now that (i), (ii) are true for the sub-trees $\mathcal{A}_1, \ldots, \mathcal{A}_m$.

Let h be a function, which is either even or odd. Denote

$$\rho(h) = \begin{cases} 1 & \text{if } h \text{ even,} \\ -1 & \text{if } h \text{ is odd.} \end{cases}$$

The function ρ is multiplicative:

$$\rho(h_1 h_2) = \rho(h_1)\rho(h_2).$$

According to the definitions of p and q and the formulas (5.28), (5.29) we have that

$$q(\lambda) = i \sin \lambda \ell \sum_{i=1}^{m} p_i(\lambda) \prod_{j \neq i} q_j(\lambda) + \cos \lambda \ell \prod_{i=1}^{m} q_i(\lambda), \qquad (5.69)$$

$$p(\lambda) = \cos \lambda \ell \sum_{i=1}^{m} p_i(\lambda) \prod_{j \neq i} q_j(\lambda) + i \sin \lambda \ell \prod_{i=1}^{m} q_i(\lambda). \qquad (5.70)$$

The hypotheses with respect to the sub-trees imply that $\rho(p_i) = -\rho(q_i)$, $i = 1, \ldots, m$. Then,

$$\rho(i \sin \lambda \ell \, p_i(\lambda) \prod_{j \neq i} q_j) = \prod_{i=1}^{m} \rho(q_i); \qquad \rho(\cos \lambda \ell \prod_{i=1}^{m} q_i) = \prod_{i=1}^{m} \rho(q_i).$$

From these relations and (5.70) we obtain

$$\rho(q) = \prod_{i=1}^{m} \rho(q_i).$$

In an analogous way it is proved that

$$\rho(p) = -\prod_{i=1}^{m} \rho(q_i).$$

From these two last equalities it holds $\rho(p) = -\rho(q)$. This proves the property (i).

We now prove (ii). If $p(\lambda_0) = q(\lambda_0) = 0$ then, from (5.69), (5.70) it follows that

$$0 = q(\lambda_0) = i \sin \lambda_0 \ell \sum_{i=1}^{m} p_i(\lambda_0) \prod_{j \neq i} q_j(\lambda_0) + \cos \lambda_0 \ell \prod_{i=1}^{m} q_i(\lambda_0),$$

$$0 = p(\lambda_0) = \cos \lambda_0 \ell \sum_{i=1}^{m} p_i(\lambda_0) \prod_{j \neq i} q_j(\lambda_0) + i \sin \lambda_0 \ell \prod_{i=1}^{m} q_i(\lambda_0).$$

This implies that

$$\sum_{i=1}^{m} p_i(\lambda_0) \prod_{j \neq i} q_j(\lambda_0) = 0, \qquad (5.71)$$

$$\prod_{i=1}^{m} q_i(\lambda_0) = 0. \tag{5.72}$$

These equalities are verified if, and only if, for some i_0

$$q_{i_0}(\lambda_0) = 0, \qquad p_{i_0}(\lambda_0) \prod_{j \neq i_0} q_j(\lambda_0) = 0$$

and this is equivalent to the fact that one of the following assertions is true:

(a) there exists $i_1 \neq i_0$ such that $q_{i_1}(\lambda_0) = 0$;

(b) $p_{i_0}(\lambda_0) = 0$.

In the first case assertion (ii) follows immediately. In (b), according to the induction assumption, there exist sub-trees of $\mathcal{A}_{i_0}$, and consequently also of $\mathcal{A}$, that verify condition (ii).

With the aid of the previous proposition it is possible to calculate how the operator Ω acts on the functions $\sin \lambda t$ and $\cos \lambda t$.

Corollary 5.21. *The following equalities are verified*

$$\Omega \sin \lambda t = \begin{cases} q(\lambda) \sin \lambda t & \text{if } q \text{ is even,} \\ -iq(\lambda) \cos \lambda t & \text{if } q \text{ is odd,} \end{cases}$$

$$\Omega \cos \lambda t = \begin{cases} q(\lambda) \cos \lambda t & \text{if } q \text{ is even,} \\ iq(\lambda) \sin \lambda t & \text{if } q \text{ is odd.} \end{cases}$$

Remark 5.22. As a consequence of the previous formulas, when q is an even function then it is real valued, while, when it is odd then iq is real valued.

5.5 Observability Results

In this section we express the inequalities from Theorem 5.13 in terms of the initial data of the solution $\bar{\phi}$. This allows us to obtain weighted observability inequalities, with explicit weights on the Fourier coefficients of the initial data of the solution. Further, we study under what conditions those weights are different from zero.

5.5.1 Weighted Observability Inequalities

As stated above, a solution $\bar{\phi}$ of (5.1)-(5.5) is expressed in terms of the initial data $\bar{\phi}_0$, $\bar{\phi}_1$ by the formula

$$\bar{\phi}(t) = \sum_{k \in \mathbb{Z}_+} \left(\phi_{0,k} \cos \lambda_k t + \frac{\phi_{1,k}}{\lambda_k} \sin \lambda_k t \right) \bar{\theta}_k, \tag{5.73}$$

where $\{\phi_{0,k}\}$, $\{\phi_{1,k}\}$ are the sequences of Fourier coefficients of $\bar{\phi}_0$, $\bar{\phi}_1$ with respect to the orthonormal basis of eigenfunctions $\{\bar{\theta}_k\}_{k \in \mathbb{Z}_+}$ and $\lambda_k = \sqrt{\mu_k}$.

Besides, the energy of the solution $\bar{\phi}$ is given by

$$\mathbf{E}_{\bar{\phi}} = \frac{1}{2} \sum_{k \in \mathbb{Z}_+} \left(\lambda_k^2 \phi_{0,k}^2 + \phi_{1,k}^2 \right). \tag{5.74}$$

The operators $\mathcal{D}_{\bar{\alpha}}$ defined in Section 5.3 are of type S. Then, according to Remark 5.5, there exist functions $d_{\bar{\alpha}}$ such that

$$\mathcal{D}_{\bar{\alpha}} e^{i\lambda t} = d_{\bar{\alpha}}(\lambda) e^{i\lambda t}.$$

In particular, when $\bar{\alpha}$ is the empty index we have $d(\lambda) \equiv 1$.

These functions, taking into account (5.37) are expressed as

$$d_{\bar{\alpha}} := \left(\prod_{i=1,\, i \neq \alpha_1}^{m} q_i \right) \left(\prod_{i=1,\, i \neq \alpha_2}^{m_{\alpha_1}} q_{\alpha_1,i} \right) \cdots \left(\prod_{i=1,\, i \neq \alpha_{k-1}}^{m_{\alpha_1,\ldots,\alpha_{k-1}}} q_{\alpha_1,\ldots,\alpha_{k-1},i} \right), \tag{5.75}$$

and then Proposition 5.20 allows to ensure that, for every $\bar{\alpha} \in \mathcal{I}$, $d_{\bar{\alpha}}$ is an even or odd function. Moreover, from Corollary 5.21 we have the equalities

$$\mathcal{D}_{\bar{\alpha}} \sin \lambda t = \begin{cases} d_{\bar{\alpha}}(\lambda) \sin \lambda t & \text{for } d_{\bar{\alpha}} \text{ even,} \\ -i d_{\bar{\alpha}}(\lambda) \cos \lambda t & \text{for } d_{\bar{\alpha}} \text{ odd,} \end{cases}$$

$$\mathcal{D}_{\bar{\alpha}} \cos \lambda t = \begin{cases} d_{\bar{\alpha}}(\lambda) \cos \lambda t & \text{for } d_{\bar{\alpha}} \text{ even,} \\ i d_{\bar{\alpha}}(\lambda) \sin \lambda t & \text{for } d_{\bar{\alpha}} \text{ odd.} \end{cases} \tag{5.76}$$

Now fix $\bar{\alpha} \in \mathcal{I}_M$ and denote $\bar{\omega} = \mathcal{D}_{\bar{\alpha}} \bar{\phi}$. The function $\bar{\omega}$ is also a solution of (5.1)-(5.5) and, from (5.73),

$$\bar{\omega}(t) = \mathcal{D}_{\bar{\alpha}} \bar{\phi}(t) = \sum_{k \in \mathbb{Z}_+} \left(\phi_{0,k} \, \mathcal{D}_{\bar{\alpha}} \cos \lambda_k t + \frac{\phi_{1,k}}{\lambda_k} \, \mathcal{D}_{\bar{\alpha}} \sin \lambda_k t \right) \bar{\theta}_k.$$

Then, from (5.76) it follows that

$$\bar{\omega}(t) = \sum_{k \in \mathbb{Z}_+} d_{\bar{\alpha}}(\lambda_k) \left(\phi_{0,k} \cos \lambda_k t + \frac{\phi_{1,k}}{\lambda_k} \sin \lambda_k t \right) \bar{\theta}_k, \qquad \text{if } d_{\bar{\alpha}} \text{ is even,}$$

$$\bar{\omega}(t) = \sum_{k \in \mathbb{Z}_+} i d_{\bar{\alpha}}(\lambda_k) \left(\phi_{0,k} \sin \lambda_k t - \frac{\phi_{1,k}}{\lambda_k} \cos \lambda_k t \right) \bar{\theta}_k, \qquad \text{if } d_{\bar{\alpha}} \text{ is odd.}$$

Thus, in both cases, the energy of $\bar{\omega}$ computed by the formula (5.74) is given by

$$\mathbf{E}_{\bar{\omega}} = \frac{1}{2} \sum_{k \in \mathbb{Z}_+} |d_{\bar{\alpha}}(\lambda_k)|^2 \left(\lambda_k^2 \phi_{0,k}^2 + \phi_{1,k}^2 \right). \tag{5.77}$$

With this, the inequality of Theorem 5.13 may be written in terms of the initial data of the solution $\bar{\phi}$ as:

$$\sum_{k \in \mathbb{Z}_+} |d_{\bar{\alpha}}(\lambda_k)|^2 \left(\lambda_k^2 \phi_{0,k}^2 + \phi_{1,k}^2 \right) \leq C \int_0^{2T_A} |F(t)|^2 dt = C \int_0^{2T_A} |\phi_x(t,0)|^2 dt.$$

$$(5.78)$$

Consequently, if we define

$$c_k = \max_{\bar{\alpha} \in \mathcal{I}_8} |d_{\bar{\alpha}}(\lambda_k)|, \qquad (5.79)$$

we obtain:

Theorem 5.23. *There exists a positive constant C, such that*

$$\sum_{k \in \mathbb{Z}_+} c_k^2 \left(\lambda_k^2 \phi_{0,k}^2 + \phi_{1,k}^2 \right) \leq C \int_0^{2T_A} |\phi_x(t,0)|^2 dt, \qquad (5.80)$$

for every solution $\bar{\phi}$ with initial data $(\bar{\phi}_0, \bar{\phi}_1) \in V \times H$.

Remark 5.24. It is easy to prove, using, e.g., formula (5.73) for the solutions, that if inequality (5.80) holds then for every $\alpha, T \in \mathbb{R}$,

$$\sum_{k \in \mathbb{Z}_+} c_k^2 \left(\lambda_k^2 \phi_{0,k}^2(T) + \phi_{1,k}^2(T) \right) \leq C \int_\alpha^{\alpha + 2T_A} |\phi_x(t,0)|^2 dt, \qquad (5.81)$$

where $\phi_{0,k}(T)$ and $\phi_{1,k}(T)$ are the Fourier coefficients of $\bar{\phi}|_{t=T}$ and $\bar{\phi}_t|_{t=T}$, respectively, in the basis $\{\bar{\theta}_k\}_{k \in \mathbb{Z}_+}$.

5.5.2 Non-degenerate Trees

In general, some of the coefficients c_k in the inequality (5.80) may vanish. That is why we consider a special class of trees for which all those numbers are different from zero.

Definition 5.25. *A tree $\mathcal{A}$ is said to be **non-degenerate** if the numbers c_k, defined for that tree by (5.79), are different from zero for every $k \in \mathbb{Z}_+$. Otherwise, the tree is said to be **degenerate**.*

The following proposition provides us with a more transparent characterization of non-degenerate trees.

Proposition 5.26. *The tree $\mathcal{A}$ is non-degenerate if and only if the spectra $\sigma_{\bar{\alpha} \diamond i}, \sigma_{\bar{\alpha} \diamond j}$ of any two sub-trees $\mathcal{A}_{\bar{\alpha} \diamond i}, \mathcal{A}_{\bar{\alpha} \diamond j}$ of $\mathcal{A}$ with common $\mathcal{O}_{\bar{\alpha}}$ root are disjoint.*

Proof. Note that the following more general fact holds: an inequality like
(5.80) with non-vanishing coefficients c_k (not necessarily given by (5.79)) is
impossible for a tree having two sub-trees with common root that share an
eigenvalue μ. Indeed, in such case, with the help of Proposition 5.15, we can
construct a non-zero solution $\bar{\phi}$ of (5.1)-(5.5) such that $\phi_x(t, 0) \equiv 0$. With
this, an inequality like (5.80) with non-trivial weights c_k would be impossible.

For the converse assertion we argue by contradiction. We will prove that
if $c_k = 0$ for some $k \in \mathbb{Z}_+$ and any two sub-trees of $\mathcal{A}$ with common root have
disjoint spectra, then $d_{\bar{\alpha}}(\lambda_k) = 0$ for any $\bar{\alpha} \in \mathcal{I}$. In particular, $d(\lambda_k) = 0$, what
would contradict the fact that $d(\lambda_k) = 1$.

Note firstly, that the property is immediate for exterior nodes, since

$$c_k \geq |d_{\bar{\alpha}}(\lambda_k)|$$

for $\bar{\alpha} \in \mathcal{I}_{\mathcal{S}}$.

For the interior ones we follow a recursive argument: if $\bar{\alpha} \in \mathcal{I}_{\mathcal{M}}$ and
$d_{\bar{\alpha} \circ \beta}(\lambda_k) = 0$ for all $\beta = 1, ..., m_{\bar{\alpha}}$ then $d_{\bar{\alpha}}(\lambda_k) = 0$.

Indeed, we have that, for every $\beta = 1, ..., m_{\bar{\alpha}}$,

$$d_{\bar{\alpha} \circ \beta} = d_{\bar{\alpha}} \prod_{i \neq \beta} q_{\bar{\alpha} \circ i}. \tag{5.82}$$

Assume that $d_{\bar{\alpha}} \neq 0$. Then (5.82) implies that

$$\prod_{i \neq 1} q_{\bar{\alpha} \circ i} = 0$$

and thus, there exists $i^* \neq 1$ such that

$$q_{\bar{\alpha} \circ i^*} = 0. \tag{5.83}$$

But then, from the equalities $d_{\bar{\alpha} \circ i^*} = 0$ and (5.82), it follows that there exists
$j^* \neq i^*$ satisfying

$$q_{\bar{\alpha} \circ j^*} = 0. \tag{5.84}$$

However, the equalities (5.83) and (5.84) ensure, according to Proposition
5.19, that $\mu_k = \lambda_k^2$ is a common eigenvalue of the sub-trees $\mathcal{A}_{\bar{\alpha} \circ i^*}$ and $\mathcal{A}_{\bar{\alpha} \circ j^*}$.
But that is impossible for the tree $\mathcal{A}$. Thus, $d_{\bar{\alpha}}(\mu_k) = 0$. This completes the
proof of the proposition.

Remark 5.27. According to the previous proposition, if the spectra of some two
sub-trees of $\mathcal{A}$ with common root have non-void intersection, inequality (5.80)
degenerates and we can not recover information on the Fourier coefficients
$\phi_{0,n}$, $\phi_{1,n}$ of the initial data of $\bar{\phi}$ from the observation of $\phi_x(t, 0)$, for those
values of n such that $c_n = 0$. However, as it has been indicated in the proof,
this is not a purely technical fact. Indeed, for degenerate trees an inequality
like (5.80) with all the coefficients c_k being different from zero, does not hold.
Thus, Theorem 5.23 is sharp in the following sense: *it provides inequality
(5.80) whenever one such inequality holds.*

Corollary 5.28 (Unique continuation property). *If the tree $\mathcal{A}$ is non-degenerate and $\bar{\phi}$ is a solution of (5.7)-(5.11) such that $\phi_x(t,0) = 0$ for almost all $t \in [0, 2L_A]$ then, $\bar{\phi} \equiv 0$.*

Remark 5.29. Combining Propositions 5.20(ii) and 5.26, we obtain an alternative characterization of non-degenerate trees: $\mathcal{A}$ is non-degenerate if, and only if,

$$|p(\lambda)|^2 + |q(\lambda)|^2 \neq 0$$

for every $\lambda \in \mathbb{R}$.

Proposition 5.30. *If the tree $\mathcal{A}$ is non-degenerate then all its eigenvalues are simple.*

Proof. If λ_k^2 is an eigenvalue of a non-degenerate tree then, according to Proposition 5.19 and Remark 5.29,

$$q(\lambda_k) = 0, \qquad p(\lambda_k) \neq 0.$$

Consequently, if $\bar{\theta}_k$ is an eigenfunction of $\mathcal{A}$ corresponding to λ_k^2, formula (5.67) gives

$$\theta^{\bar{\alpha}}(0) = \frac{\hat{k}(\lambda_k)}{i\lambda_k p(\lambda_k)}\theta_x(0), \qquad \theta_x^{\bar{\alpha}}(0) = \frac{-l(\lambda_k)}{p(\lambda_k)}\theta_x(0).$$

Thus, $\bar{\theta}_k$ is determined, up to the constant factor $\theta_x(0)$, in a unique way.

Remark 5.31. Let $(\tilde{\mu}_k)_{n \in \mathbb{Z}_+}$ be the strictly increasing sequence of the eigenvalues μ_k of a tree without taking into account their multiplicity. In Chapter 6 we will prove that $\tilde{\mu}_k$ verifies $\mu_k^N \leq \tilde{\mu}_k \leq \mu_k^D$, for $k \in \mathbb{Z}_+$, where $\{\mu_k^D\}_{k \in \mathbb{Z}_+}$ and $\{\mu_k^N\}_{k \in \mathbb{Z}_+}$ are the ordered sequences formed by the distinct eigenvalues of the strings entering in the network with Dirichlet or Neumann homogeneous boundary conditions, respectively. This fact will allow to prove that an inequality of type (5.80) is impossible for $T < 2L_A$ (see Theorem 6.5). Moreover, in this case the system (5.1)-(5.5) is not approximately controllable, and then, is not spectrally controllable either.

5.5.3 On the Set of Non-degenerate Trees

Now we give some information on the size of the set of degenerate trees. It turns out that almost all trees with the same topological structure are non-degenerate in the sense of a measure, defined in a natural way on the set of trees with that structure.

Let us be more precise. We shall say that two trees are *topologically equivalent* if their edges can be numbered with the same set of multi-indices. This means that they may differ only in the lengths of their edges. In particular, two equivalent trees have the same number of edges and vertices. The classes of topologically equivalent trees are called *topological configurations*.

Fix a topological configuration Σ with d edges. We assume that in the set of indices $\mathcal{I}$ for the elements of the trees belonging to Σ, a criterion of ordering has been defined and use the notation $< \mathcal{A} >$ for the corresponding ordered set of the lengths of the edges of $\mathcal{A} \in \Sigma$.

Then Σ may be identified with $(\mathbb{R}_+)^d$ by means of the canonical mapping $\pi : \Sigma \to \mathbb{R}^d$ defined by

$$\pi(\mathcal{A}) = < \mathcal{A} > \in \mathbb{R}^d.$$

Let μ_Σ be the measure induced in Σ by the Lebesgue measure of $\mathbb{R}^d$ through the mapping π. That is, if $B \subset \Sigma$ then

$$\mu_\Sigma(B) = m_d(\pi(B)),$$

where m_d is the usual Lebesgue measure in $\mathbb{R}^d$.

The following holds:

Proposition 5.32. *Given a topological configuration Σ, almost every tree (in the sense of the measure μ_Σ) with that topological configuration is non-degenerate.*

Proof. Let $D_{\bar{\alpha}}^{i,j} \subset \Sigma$ denote the set of those trees $\mathcal{A}$, such that its sub-trees $\mathcal{A}_{\bar{\alpha}\circ i}$ and $\mathcal{A}_{\bar{\alpha}\circ j}$ are non-degenerate and have a common eigenvalue. Then the set $\Sigma_{deg} \subset \Sigma$ of degenerate trees may be decomposed as

$$\Sigma_{deg} = \bigcup_{\bar{\alpha} \in \mathcal{I}_M} \bigcup_{i,j=1 \; i \neq j}^{m_{\bar{\alpha}}} D_{\bar{\alpha}}^{i,j}. \tag{5.85}$$

We will prove that $\mu_\Sigma(D_{\bar{\alpha}}^{i,j}) = 0$, for every $\bar{\alpha} \in \mathcal{I}_M$, $i,j = 1, ..., m_{\bar{\alpha}}$, $i \neq j$. This fact, in view of (5.85), will imply $\mu_\Sigma(\Sigma_{deg}) = 0$. In what follows we consider that $\bar{\alpha}$, i and j are fixed.

The idea of the proof is simple. We fix a tree $\mathcal{B}$ having the structure[2] of $\mathcal{A}_{\bar{\alpha}\circ i} \vee \mathcal{A}_{\bar{\alpha}\circ j}$ (defined as in Remark 5.16) and extend it adding edges to a tree $\mathcal{A} \in D_{\bar{\alpha}}^{i,j}$. According to Remark 5.16, that is equivalent to choosing the node $\mathcal{O}_{\bar{\alpha}}$ of $\mathcal{A} \in \Sigma$ in a point of a string of $\mathcal{B}$ (precisely, of that string where it should be located to agree with the structure of Σ) where some eigenfunction of $\mathcal{B}$ vanishes. Once $\mathcal{O}_{\bar{\alpha}}$ has been chosen, the lengths of the remaining strings of $\mathcal{A}$ may be taken arbitrarily.

Observe that we may assume that no (non-identically zero) eigenfunction of $\mathcal{B}$ vanishes identically on the string that contains $\mathcal{O}_{\bar{\alpha}}$, since, otherwise, one of the sub-trees of $\mathcal{A}_{\bar{\alpha}\circ i}$ or $\mathcal{A}_{\bar{\alpha}\circ j}$ of the tree $\mathcal{A}$, obtained with this procedure, would be degenerate and thus, $\mathcal{A} \notin D_{\bar{\alpha}}^{i,j}$. This assumption implies that all the eigenfunctions of $\mathcal{B}$ are simple and then the node $\mathcal{O}_{\bar{\alpha}}$ should be chosen in a set of points, which is at most denumerable.

Thus, we have obtained, after some re-ordering if needed, that the set $\pi(D_{\bar{\alpha}}^{i,j})$ is contained in a set of the form

---

[2] That is, $\mathcal{B}$ is topologically equivalent to $\mathcal{A}_{\bar{\alpha}\circ i} \vee \mathcal{A}_{\bar{\alpha}\circ j}$.

$$\left\{ (h_1, h_2, ..., h_d) \in (\mathbb{R}_+)^d : h_1 + h_2 = h, \; h_1 \in \mathbf{N}(h, h_3, ..., h_d) \right\}, \qquad (5.86)$$

where $\mathbf{N}(h, h_3, ..., h_d)$ is a denumerable set depending on h and $h_3, ..., h_d$.

It is easy to see, using, e.g., the Fubbini's theorem, that a set defined by (5.86) has zero d-dimensional Lebesgue measure. Thus, the same is true for $\pi(D_{\bar{\alpha}}^{i,j})$ and then $\mu_{\Sigma}(D_{\bar{\alpha}}^{i,j}) = 0$. This completes the proof.

Corollary 5.33. *The set* $\Sigma \setminus \Sigma_{deg}$ *of non-degenerate trees is dense in* Σ *endowed with the metrics induced in* Σ *by the usual metrics of* $\mathbb{R}^d$ *through* π.

Remark 5.34. The set Σ_{deg}, even though is small in the sense of μ_{Σ}, is dense in Σ. Indeed, it suffices to see that, if two edges of a tree with rationally dependent lengths have a common vertex and their other vertices are exterior then the tree is degenerate.

5.6 Consequences Concerning Controllability

Gathering the results of the previous sections we obtain the following characterization of the controllability properties of trees.

Theorem 5.35. *Let* $\mathcal{A}$ *be a tree and* $T > 0$. *Then,*
a) If $T \geq 2L_{\mathcal{A}}$ *the following properties are equivalent:*

1) system (5.1)-(5.6) is spectrally controllable in time T;
2) system (5.1)-(5.6) is approximately controllable in time T;
3) $\mathcal{A}$ *is non-degenerate;*
4) any two sub-trees of $\mathcal{A}$ *with common root have disjoint spectra.*

Moreover, when they are true, all the initial states of the space $\mathcal{W}$, *defined by*

$$\mathcal{W} = \left\{ (\bar{u}_0, \bar{u}_1) \in V \times H' : \sum_{n \in \mathbb{N}} \frac{1}{c_n^2} \left(|u_n^0|^2 + \frac{1}{\mu_n} |u_n^1|^2 \right) < \infty \right\},$$

where the weights c_n *are given by (5.79), are controllable in time* T. *Besides, these properties are true for almost all trees, topologically equivalent to* $\mathcal{A}$.

b) If $T < 2L_{\mathcal{A}}$ *the spectral controllability property fails, independently on whether the tree is degenerate or not.*

Proof. The assertion a) follows from the following implications:

1) $\Rightarrow$ 2). This is a general fact, since $Z \times Z$ is dense in $H \times V'$, the spectral controllability is a particular case of approximate controllability.

2) $\Rightarrow$ 3). It follows from Proposition 5.26 and Remark 5.27, that, for degenerate trees, there exist non-zero eigenfunctions, which vanish identically on the string that contains the root. Those eigenfunctions do not satisfy the unique continuation property from the root, whatever the time T is. Consequently, approximate controllability fails as well.

3) $\Rightarrow$ 4). This has been proved in Proposition 5.26.

4) $\Rightarrow$ 1). It is also a consequence of Proposition 5.26. If any two sub-trees of $\mathcal{A}$ with common root have disjoint spectra, there exists a constant C, such that

$$\sum_{n \in \mathbb{N}} c_n^2 \left(\lambda_n^2 \phi_{0,n}^2 + \phi_{1,n}^2 \right) \leq C \int_0^{2L_\mathcal{A}} |\phi_x(t,0)|^2 dt,$$

for every solution $\bar{\phi}$ of (5.7)-(5.12) with initial state $(\bar{\phi}_0, \bar{\phi}_1) \in V \times H$, where all the coefficients c_n are different from zero. This implies that the space of initial states

$$\mathcal{W} = \left\{ (\bar{u}_0, \bar{u}_1) \in V \times H' : \sum_{n \in \mathbb{N}} \frac{1}{c_n^2} \left(|u_n^0|^2 + \frac{1}{\mu_n} |u_n^1|^2 \right) < \infty \right\}$$

is controllable in any time $T \geq 2L_\mathcal{A}$. In particular, the space $Z \times Z$ will be controllable.

The assertion b) is true in the general case of arbitrary networks, whose structure is not necessarily a tree. This fact will be proved in Theorem 6.5.

5.7 Simultaneous Observability and Controllability of Networks

The results of the previous sections allow to consider the one-node control problem for several (a finite number) of tree-shaped networks when the same control function is used to control all of them, i.e., when they are controlled simultaneously.

Let $\mathcal{A}^1, ..., \mathcal{A}^R$ be the trees associated to the controlled networks. For the elements of the network whose graph is $\mathcal{A}^r$ we will use the same notations as in the preceding sections but adding the superscript r to them. Thus, the solution of (5.1)-(5.6) for the tree $\mathcal{A}^r$ (in what follows we shall briefly refer to this problem as $(5.1)_r$-$(5.6)_r$) is denoted by $\bar{u}^r$ and the spaces V and H constructed for that tree by V^r and H^r.

We define the space

$$\mathbf{W} = \prod_{r=1}^{R} [V^r \times H^r],$$

endowed with the product Hilbert structure. The elements of $\mathbf{W}$ are called *simultaneous states*.

We shall say that the simultaneous state $\bar{w} \in \mathbf{W}$, is *controllable in time T* if it is possible to find a control function $v \in L^2(0,T)$ such that the solutions $\bar{u}^r$ of $(5.1)_r$-$(5.6)_r$ with initial states $(\bar{u}_0^r, \bar{u}_1^r)$ (the components of $\bar{w}$) and $v^r = v$ verify

$$\bar{u}^r(T,x) = \bar{u}_t^r(T,x) = \bar{0},$$

for every $r = 1, ..., R$.

We underline that the control v applied to each of the trees is the same. Consequently, the problem under consideration consists in controlling R trees by means of the same control applied at a common node.

Once again using HUM, the problem of characterizing the controllable simultaneous states is reduced to the study of observability inequalities for the corresponding homogeneous system. Indeed, assume that there exist non-zero numbers c_n^r, $n \in \mathbb{Z}_+$, $r = 1, ..., R$, such that for every k the inequality

$$\int_0^T |\sum_{r=1}^R \phi_x^r(0,t)|^2 dt \geq \sum_{n \in \mathbb{Z}_+} (c_n^k)^2 \left(\mu_n^k |\phi_{0,n}^k|^2 + |\phi_{1,n}^k|^2 \right), \qquad (5.87)$$

holds for all solutions of the adjoint system. Note that the adjoint system is simply the superposition of the homogeneous wave equation $(5.7)_r$-$(5.12)_r$ with Dirichlet boundary conditions in each sub-tree $\mathcal{A}^r$, that we denote by $\bar{\phi}^r$. Observe, in particular, that the adjoint system is decoupled.

Here $\{\phi_{0,n}^r\}$ and $\{\phi_{1,n}^r\}$ are the sequences of Fourier coefficients of the components $\bar{\phi}_0^r$ and $\bar{\phi}_1^r$ of the initial state in the bases $\{\bar{\theta}_n^r\}$ of H^r, respectively, and $\bar{\phi}^r$ is the solution of $(5.7)_r$-$(5.12)_r$.

Define the sets

$$\mathcal{W}^r = \{(\bar{u}_0^r, \bar{u}_1^r) \in V^r \times (H^r)' : \|(\bar{u}_0^r, \bar{u}_1^r)\|_r < \infty\}, \qquad (5.88)$$

where

$$\|(\bar{u}_0^r, \bar{u}_1^r)\|_r^2 := \sum_{n \in \mathbb{Z}_+} \frac{1}{(c_n^r)^2} \left(|u_{0,n}^r|^2 + \frac{1}{\mu_n^r} |u_{1,n}^r|^2 \right). \qquad (5.89)$$

Then, if the inequalities (5.87) hold for the solutions of the homogeneous system all the initial simultaneous states $\bar{w} \in \mathcal{W} = \prod_{i=1}^r \mathcal{W}^r$ are controllable in time T.

In particular, the initial simultaneous states $\bar{w} \in \prod_{1=1}^r Z^r \times Z^r$ are controllable (recall that Z^r is the space of all finite linear combinations of the eigenfunctions $\bar{\theta}_n^r$). In this case, the networks are said to be *simultaneously spectrally controllable*.

Moreover, the set of controllable simultaneous states in time T is dense in **W** (when that holds the networks are said to be *simultaneously approximately controllable in time T*) if and only if the following unique continuation property holds for the adjoint system:

$$\sum_{r=1}^R \phi_x^r(0,t) = 0 \quad a. \ e. \ t \in (0,T) \Rightarrow (\bar{\phi}_0^r, \bar{\phi}_1^r) = \bar{0}, \quad \forall r = 1, ..., R. \qquad (5.90)$$

It is clear that, if a simultaneous state is controllable, then each of its components is also controllable for the corresponding network. This implies that the approximate controllability of each indvidual sub-tree is a necessary condition for the simultaneous approximate controllability. In other words, each sub-tree being non-degenerate is a necessary condition for the simultaneous approximate controllability.

On the other hand, if two of the trees, say $\mathcal{A}^1$ and $\mathcal{A}^2$, have a common eigenvalue then, using the pasting procedure described in the proof of Proposition 5.15, we can construct non-zero solutions of $(5.7)_r$-$(5.12)_r$, $r = 1, 2$, such that

$$\phi_x^1(t, 0) + \phi_x^2(t, 0) = 0, \qquad t \in \mathbb{R}.$$

Extending these solutions by zero to all the remaining trees $\mathcal{A}^r$, $r = 3, ..., R$, we obtain a simultaneous solution $\bar{\phi}$ of the adjoint system for which inequalities (5.87) are impossible and, moreover, for which the unique continuation property (5.90) fails.

Thus, the conditions that the trees $\mathcal{A}^r$, $r = 1, ..., R$, are non-degenerate and their spectra are pairwise disjoint are necessary for the simultaneous approximate controllability, and, consequently, for the spectral controllability as well. As we shall see, these conditions are also sufficient.

Put $T^* = \sum_{i=1}^r L^i$. For every $k = 1, ..., R$ we define the operator

$$\widehat{\mathcal{Q}}_k := \prod_{r=1,\, r\neq k}^R \mathcal{Q}^r,$$

where $\mathcal{Q}^r$ is the operator $\mathcal{Q}$ for the tree $\mathcal{A}^r$. Note that $\widehat{\mathcal{Q}}_k$ is an S-operator with $s(\widehat{\mathcal{Q}}_k) = T^* - L^k$.

Let $\widehat{q}_k$ be the function associated to $\widehat{\mathcal{Q}}_k$ according to Remark 5.5. Then

$$\widehat{q}_k = \prod_{r=1,\, r\neq k}^R q^r, \qquad (5.91)$$

where q^r is the function corresponding to $\mathcal{Q}^r$.

Proposition 5.36. *If for a given r there exist numbers c_n, $n \in \mathbb{Z}_+$, such that every solution $\bar{\phi}$ of $(5.7)_r$-$(5.12)_r$ with initial state*

$$(\bar{\phi}_0^r, \bar{\phi}_1^r) = \Big(\sum_{n\in\mathbb{Z}_+} \phi_{0,n}^r \bar{\theta}_n^r, \sum_{n\in\mathbb{Z}_+} \phi_{1,n}^r \bar{\theta}_n^r \Big) \in V^r \times H^r$$

satisfies

$$\int_0^{2L_r} |\phi_x^r(0, t)|^2 dt \geq \sum_{n\in\mathbb{Z}_+} c_n^2 \left(\mu_n^r |\phi_{0,n}^r|^2 + |\phi_{1,n}^r|^2 \right), \qquad (5.92)$$

then

$$\int_0^{2T^*} \Big| \sum_{r=1}^R \phi_x^r(0, t) \Big|^2 dt \geq \sum_{n\in\mathbb{Z}_+} c_n^2 |\widehat{q}_k(\lambda_n^k)|^2 \left(\mu_n^k |\phi_{0,n}^k|^2 + |\phi_{1,n}^k|^2 \right), \qquad (5.93)$$

for every $k = 1, ..., R$.

Proof. As $\widehat{\mathcal{Q}}_k$ is an S-operator with $s(\widehat{\mathcal{Q}}_k) = T^* - L^k$, using Proposition 5.3(i) we get

$$\int_0^{2T^*} |\sum_{r=1}^{R} \phi_x^i(0,t)|^2 dt \geq \int_{T^*-L^k}^{T^*+L^k} |\widehat{\mathcal{Q}}_k \sum_{r=1}^{R} \phi_x^i(0,t)|^2 dt. \qquad (5.94)$$

But, as $\mathcal{Q}^r \phi_x^r(0,t) = 0$, then $\widehat{\mathcal{Q}}_k \bar{\phi}^r = 0$ if $r \neq k$. Thus, inequality (5.94) becomes

$$\int_0^{2T^*} |\sum_{r=1}^{R} \phi_x^i(0,t)|^2 dt \geq \int_{T^*-L^k}^{T^*+L^k} |\widehat{\mathcal{Q}}_k \phi_x^k(0,t)|^2 dt. \qquad (5.95)$$

Now we consider the function $\bar{\omega} = \widehat{\mathcal{Q}}_k \bar{\phi}$. As $\bar{\omega}$ is clearly a solution of $(5.1)_k$-$(5.5)_k$, then, according to (5.92) and Remark 5.81 it holds

$$\int_{T^*-L^k}^{T^*+L^k} |\omega_x(0,t)|^2 dt \geq \sum_{n \in \mathbb{Z}_+} c_n^2 \left(\mu_n^k \omega_{0,n}^2 + \omega_{1,n}^2 \right). \qquad (5.96)$$

On the other hand, it is simple to prove that the Fourier coefficients of the initial data of $\bar{\phi}$ and $\bar{\omega}$ are related by

$$\mu_n^k \omega_{0,n}^2 + \omega_{1,n}^2 = q_k^2(\lambda_n^k) \left(\mu_n^k \phi_{0,n}^2 + \phi_{1,n}^2 \right). \qquad (5.97)$$

Finally, combining (5.95)-(5.97) and the fact that $\omega_x(0,t) = \widehat{\mathcal{Q}}_k \phi_x^k(0,t)$, the inequality (5.93) is obtained.

Now, if the trees $\mathcal{A}^1,..., \mathcal{A}^R$ are non-degenerate for every $r = 1,..., R$, inequalities (5.92) hold with non-zero coefficients c_n (depending on r), which are explicitly computed by formulas (5.79). Therefore, according to Proposition 5.36, we shall also have inequalities (5.87) with explicitly computed coefficients

$$c_n^r = |\widehat{q}_r(\lambda_n^r)| c_n,$$

which are all different from zero whenever the spectra of any two of the trees $\mathcal{A}^r$ are disjoint, since $\widehat{q}_r(\lambda_n^r) \neq 0$ for all $r = 1,..., R$ and $n \in \mathbb{Z}_+$. Indeed, if $\widehat{q}_r(\lambda_n^r) = 0$ for some r and n then equality (5.91) would imply that $q^i(\lambda_n^r) = 0$ for some $i \neq r$ and thus, from Proposition (5.19), μ_n^r would be a common eigenvalue of the trees $\mathcal{A}^r$ and $\mathcal{A}^i$.

Consequently, under those assumptions, we are able to construct a space

$$\mathcal{W} = \prod_{r=1}^{R} \mathcal{W}^r,$$

where $\mathcal{W}^r$ are defined by (5.88)-(5.89), of controllable simultaneous states in time $2T^*$. In particular,

Corollary 5.37. *The trees $\mathcal{A}^1,..., \mathcal{A}^R$ are simultaneously spectrally controllable in some time T (and then in time $2T^*$), if and only if they are spectrally controllable and their spectra are pairwise disjoint.*

5.8 Examples

5.8.1 The Star-Shaped Network with n Strings

In the framework of the study of the controllability of networks of strings from an exterior node, the star-shaped network with n strings constitutes the simplest example. Obviously, the number of strings involved in the star-shaped network is arbitrary.

The star-shaped network with n strings is formed by n strings connected at one point. When $n = 3$, this network is the three string network studied in Chapter 4.

Let us call $\mathcal{A}$ the star-shaped graph that supports the network. Following the numbering criterion introduced in Section 1 for trees, we will denote by $\mathcal{R}$ the controlled node and by $\mathcal{O}$ the interior one where the strings are coupled. The controlled string will be denoted by $\mathbf{e}$ and its length by ℓ. The remaining $n-1$ exterior nodes are denoted by $\mathcal{O}_i$, $i = 1, ..., n-1$, the string that contains $\mathcal{O}_i$ by $\mathbf{e}_i$ and its length by ℓ_i.

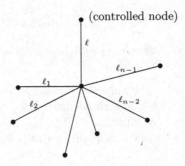

Fig. 5.3. Star-shaped network with n strings

The only sub-trees of $\mathcal{A}$ are the strings $\mathbf{e}_i$, $i = 1, ..., n-1$: $\mathcal{A}_i = \{\mathbf{e}_i\}$. Therefore, the spectra σ_i of the sub-trees coincide with the eigenvalues of the homogeneous Dirichlet problem for a string and are given by

$$\sigma_i = \left\{ \left(\frac{k\pi}{\ell_i} \right)^2 : \quad k \in \mathbb{Z} \right\}, \qquad i = 1, ..., n-1.$$

The non-degeneracy condition of $\mathcal{A}$ is $\sigma_i \cap \sigma_j = \emptyset$ for every pair i, j with $i \neq j$. This means

$$\frac{k\pi}{\ell_i} \neq \frac{m\pi}{\ell_j}$$

for all $k, m \in \mathbb{Z}$, which is equivalent to the fact that ℓ_i/ℓ_j is irrational.

Applying Theorem 5.35 we conclude that if $L_{\mathcal{A}}$ is the sum of the lengths of all the strings of the network it holds:

Corollary 5.38. *The star-shaped network with n strings is approximately controllable in some time $T \geq 2L_{\mathcal{A}}$ (and then spectrally controllable in time $T = 2L_{\mathcal{A}}$) if, and only if, the ratio of any two of the lengths of the uncontrolled strings is an irrational number.*

Besides, when the non-degeneracy condition is fulfilled, all the initial states $(\bar{u}_0, \bar{u}_1) \in V' \times H$ satisfying

$$\sum_{k \in \mathbb{N}} \frac{1}{c_k^2} u_{0,k}^2 < \infty, \quad \sum_{k \in \mathbb{N}} \frac{1}{c_k^2 \mu_k} u_{1,k}^2 < \infty, \tag{5.98}$$

are controllable in time $T = 2L_{\mathcal{A}}$. Recall than in (5.98) $\mu_k = \lambda_k^2$ are the eigenvalues of the network and the coefficients c_k are defined by (5.79):

$$c_k = \max_{i=1,\dots,n-1} \prod_{j \neq i} |q_j(\lambda_k)|,$$

where q_j is the function associated to the operator $\mathcal{Q}_j$ for the sub-tree $\mathcal{A}_j$.

But, as $\mathcal{A}_j$ coincides with the string $\mathbf{e}_j$, the operator $\mathcal{Q}_j$ coincides with ℓ_j^- (see Subsection 5.2.1, where the operators $\mathcal{P}$ and $\mathcal{Q}$ are computed for a string) and then, from Remark 5.5,

$$q_j(\lambda_k) = i \sin \lambda_k \ell_j.$$

In conclusion,

$$c_k = \max_{i=1,\dots,n-1} \prod_{j \neq i} |\sin \lambda_k \ell_j|.$$

In Appendix A we pay special attention to the function

$$\mathbf{a}(\lambda, \ell_1, \dots, \ell_{n-1}) := \sum_{i=1}^{n-1} \prod_{j \neq i} |\sin \lambda \ell_j|.$$

There we provide conditions on the values of $\ell_1, \dots, \ell_{n-1}$ such that, for every $\lambda \in \mathbb{R}$, an inequality of the type

$$\mathbf{a}(\lambda, \ell_1, \dots, \ell_{n-1}) \geq C \lambda^\alpha$$

is satisfied. These conditions involve certain sets $\mathbf{B}_\varepsilon$, $\varepsilon > 0$, which are defined in Appendix A, p. 208, where in addition, some conditions on the lengths, called *conditions* (S), are introduced.

As, obviously, $n c_k \geq \mathbf{a}(\lambda_k, \ell_1, \dots, \ell_{n-1})$ then, applying Corollary A.10 we obtain

Corollary 5.39. *If the numbers $\ell_1, \dots, \ell_{n-1}$ are such that for all values $i, j = 1, \dots, n-1$, $i \neq j$, the ratios ℓ_i / ℓ_j belong to $\mathbf{B}_\varepsilon$ then, there exists a constant $C_\varepsilon > 0$ such that*

$$c_k \geq \frac{C_\varepsilon}{\lambda^{n-2+\varepsilon}}, \qquad k \in \mathbb{N}.$$

Therefore, all the initial states $(\bar{u}_0, \bar{u}_1) \in V^{n-2+\varepsilon} \times V^{n-3+\varepsilon}$ are controllable in time $T = 2L_{\mathcal{A}}$.

Under more restrictive assumptions on the lengths of uncontrolled strings, it is possible to guarantee the existence of a larger subspace of controllable initial states:

Corollary 5.40. *If the numbers $\ell_1, ..., \ell_{n-1}$ verify the conditions (S) then, for every $\varepsilon > 0$, there exists a constant $C_\varepsilon > 0$ such that*

$$c_k \geq \frac{C_\varepsilon}{\lambda^{1+\varepsilon}}, \qquad k \in \mathbb{N}.$$

Therefore, all the initial states $(\bar{u}_0, \bar{u}_1) \in V^{1+\varepsilon} \times V^\varepsilon$ are controllable in time $T = 2L_\mathcal{A}$.

Remark 5.41. When $n = 3$ the results of Corollaries 5.39 and 5.40 coincide with Corollary 4.30 relative to the three string network.

5.8.2 Simultaneous Control of n Strings

This problem is quite similar to the previous one, though in fact it is simpler. It consists in controlling n strings $\mathbf{e}_1, ..., \mathbf{e}_n$ of lengths $\ell_1, ..., \ell_n$, which are not coupled; we just use the same function to control all the strings. This is the simplest problem of simultaneous control of an arbitrary number of tree-shaped networks in the sense of Section 5.7. Let us note that the case $n = 2$ has been already studied in Section 4.2 of Chapter 4. As it was pointed out there, this is the problem studied in [12], [10], [17], [15], [16], with the help of Theorem 3.32. In [35] this problem was solved using the technique we describe here.

The controlled system is

$$\begin{cases} u_{tt}^i - u_{xx}^i = 0 & (t, x) \in \mathbb{R} \times [0, \ell_i], \\ u^i(t, \ell_i) = 0, \quad u^i(t, 0) = v(t) & t \in \mathbb{R}, \\ u^i(0, x) = u_0^i(x), \quad u_t^i(0, x) = u_1^i(x) & x \in [0, \ell_i], \end{cases} \tag{5.99}$$

for $i = 1, ..., n$.

Let us observe that it may be viewed as a star-shaped network with n controlled strings, the control being applied at the interior node, that is, at the coupling point.

According to Corollary 5.37 of Section 5.7, the n strings are simultaneously spectrally controllable in some time T (and then also in time $T_0 = 2 \sum_{i=1}^n \ell_i$) if, and only if, the spectra of any two strings are disjoint. This is equivalent to the fact that all the ratios ℓ_i / ℓ_j with $i \neq j$ are irrational numbers.

It is possible to obtain additional information directly from Proposition 5.36. In this case, the controlled trees are strings: $\mathcal{A}_i = \{\mathbf{e}_i\}$. Then we have $\mathcal{Q}_i = \ell_i^-$ and therefore

$$\mathcal{Q}_i = \ell_i^-, \qquad \widehat{\mathcal{Q}}_i = \prod_{j \neq i} \mathcal{Q}_j = \prod_{j \neq i} \ell_j^-, \qquad |\widehat{q}_i(\lambda)| = \prod_{j \neq i} |\sin(\lambda \ell_j)|.$$

Besides, the eigenvalues (μ_k^i) and eigenfunctions (θ_k^i) of each $\mathcal{A}_i$ may be explicitly computed:

$$\mu_k^i = \left(\frac{k\pi}{\ell_i}\right)^2, \qquad \theta_k^i(x) = \sqrt{\frac{2}{\ell_i}}\sin(\frac{k\pi}{\ell_i}x).$$

On the other hand, if $(\phi_{0,k}^i), (\phi_{1,k}^i)$ denote the sequences of Fourier coefficients of the initial state $(\bar\phi_0^i, \bar\phi_1^i)$ of the string $\mathbf{e}_i$ in the basis (θ_k^i), then

$$\int_0^{2\ell_i}\left|\phi_x^i(t,0)\right|^2 \geq 4\sum_{k\in\mathbb{N}}\left(\mu_k^i(\phi_{0,k}^i)^2 + (\phi_{1,k}^i)^2\right), \tag{5.100}$$

for every solution

$$\phi_{tt}^i - \phi_{xx}^i = 0, \qquad \phi^i(t,0) = \phi^i(t,\ell_i) = 0,$$

with initial states $(\bar\phi_0^i, \bar\phi_1^i) \in Z_i \times Z_i$ (this is the observability inequality of a string from one of its extremes, see Proposition 3.1).

Applying Proposition 5.36 to the inequalities (5.100) we deduce

$$\int_0^{T^*}\left|\sum_{i=1}^n \phi_x^i(t,0)\right|^2 dt \geq 4\sum_{k\in\mathbb{N}}\left|\widehat{q}_i(\lambda_k^i)\right|^2 \left(\mu_k^i(\phi_{0,k}^i)^2 + (\phi_{1,k}^i)^2\right),$$

for every $i = 1, ..., n$.

These are the observability inequalities associated to the control of system (5.99). As a consequence of this we get that, if for each $i = 1, ..., n$, the simultaneous initial state (u_0^i, u_1^i), $i = 1, ..., n$, satisfies

$$\sum_{k\in\mathbb{N}}\frac{(u_{0,k}^i)^2}{\left|\widehat{q}_i(\lambda_k^i)\right|^2} < \infty, \qquad \sum_{k\in\mathbb{N}}\frac{(u_{1,k}^i)^2}{\left|\widehat{q}_i(\lambda_k^i)\right|^2 \mu_k^i} < \infty,$$

then, that state is controllable in time T^*.

The weights $\left|\widehat{q}_i(\lambda_k^i)\right|$ may be easily estimated:

$$\left|\widehat{q}_i(\lambda_k^i)\right| = \prod_{j\neq i}\left|\sin(\lambda_k^i\ell_j)\right| = \prod_{j\neq i}\left|\sin(k\pi\frac{\ell_j}{\ell_i})\right| \geq C\prod_{j\neq i}\left|\left|\left|k\frac{\ell_j}{\ell_i}\right|\right|\right|.$$

(Recall that $|||\eta|||$ denotes the distance from η to $\mathbb{Z}$.)

Then we obtain, in account of the results on Diophantine Approximation included in Appendix A, conditions that allow to identify subspaces of simultaneous controllable states in time $T^* = 2\sum_{i=1}^n \ell_i$:

Corollary 5.42. *If the numbers $\ell_1, ..., \ell_n$ are such that for all the values $i, j = 1, ..., n$, $i \neq j$, the ratios ℓ_i/ℓ_j belong to $\mathbf{B}_\varepsilon$ then there exists a constant $C_\varepsilon > 0$ such that*

$$|\widehat{q_i}(\lambda_k^i)| \geq \frac{C_\varepsilon}{(\lambda_k^i)^{n-1+\varepsilon}}, \qquad k \in \mathbb{N}.$$

Therefore, the space of controllable simultaneous initial states in time $T^* = 2\sum_{i=1}^n \ell_i$ *contains all those simultaneous states that verify*

$$(u_0^i, u_1^i) \in V_i^{1+\varepsilon} \times V_i^\varepsilon,$$

for every $i = 1, ..., n$, *where* V_i^α *is defined as in (2.27) for the string* $\mathbf{e}_i$, *that is,*

$$V_i^\alpha = \left\{ \varphi = \sum_{k \in \mathbb{N}} \varphi_k \sin(k\pi x/\ell_i) : \sum_{k \in \mathbb{N}} k^{2\alpha} |\varphi_k|^2 < \infty \right\}.$$

Corollary 5.43. *If the lengths* $\ell_1, ..., \ell_n$ *verify the conditions (S) then, for every* $\varepsilon > 0$ *there exists a constant* $C_\varepsilon > 0$ *such that*

$$|\widehat{q_i}(\lambda_k^i)| \geq \frac{C_\varepsilon}{(\lambda_k^i)^{1+\varepsilon}}, \qquad k \in \mathbb{N}.$$

Therefore, the space of controllable simultaneous initial states in time $T^* = 2\sum_{i=1}^n \ell_i$ *contains all those simultaneous states that verify*

$$(u_0^i, u_1^i) \in V_i^{1+\varepsilon} \times V_i^\varepsilon,$$

for every $i = 1, ..., n$.

Remark 5.44. The problem of simultaneous control of n strings may be successfully studied with the aid of the method of moments. It suffices to note that the function

$$F(z) = \prod_{i=1}^n \sin z\ell_i$$

is a generating function of the increasing sequence (σ_m) formed by the numbers λ_k^i, $i = 1, ..., n$, $k \in \mathbb{N}$ (the positive square root of the eigenvalues of the strings).

The function F is bounded and of exponential type $A = \sum_{i=1}^n \ell_i$. Besides,

$$|F'(\lambda_k^i)| = \prod_{j \neq i} \left| \sin(k\pi \frac{\ell_j}{\ell_i}) \right|.$$

This allows to obtain results similar to those of Corollaries 5.42 and 5.43.

5.8.3 A Non Star-Shaped Tree

Now let us consider a tree $\mathcal{A}$, which is not star-shaped, having a very simple structure as shown in Figure 5.4. We will assume in addition that $\ell_{1,2} = \ell_2$.

This tree contains four sub-trees. Two of them

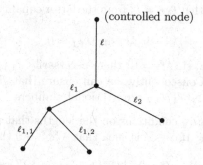

Fig. 5.4. A tree which is not star-shaped

$$\mathcal{A}_1 = \{\mathbf{e}_1, \mathbf{e}_{1,1}, \mathbf{e}_{1,2}\}, \qquad \mathcal{A}_2 = \{\mathbf{e}_2\},$$

have the common root $\mathcal{O}$. The other two

$$\mathcal{A}_{1,1} = \{\mathbf{e}_{1,1}\}, \qquad \mathcal{A}_{1,2} = \{\mathbf{e}_{1,2}\},$$

have the common root $\mathcal{O}_1$ and are constituted by one single string.
The operators $\mathcal{Q}$ corresponding to these sub-trees are

$$\mathcal{Q}_2 = \ell_2^-, \qquad \mathcal{Q}_{1,1} = \ell_{1,1}^-, \qquad \mathcal{Q}_{1,2} = \ell_{1,2}^-,$$

$$\mathcal{Q}_1 = \left(\ell_1^+ \ell_{1,1}^- \ell_{1,2}^- + \ell_1^- \ell_{1,1}^+ \ell_{1,2}^- + \ell_1^- \ell_{1,1}^- \ell_{1,2}^+\right).$$

The first three operators are obtained immediately, since the corresponding
sub-trees are strings. The operator corresponding to $\mathcal{A}_1$, is the operator $\mathcal{Q}$ for
a three string network and has been constructed in Chapter 4.
The functions associated to these operators are

$$q_2(\lambda) = i\sin\lambda\ell_2, \qquad q_{1,1}(\lambda) = i\sin\lambda\ell_{1,1}, \qquad q_{1,2}(\lambda) = i\sin\lambda\ell_{1,2},$$

$$q_1(\lambda) = -(\cos\lambda\ell_1 \sin\lambda\ell_{1,1} \sin\lambda\ell_{1,2} + \sin\lambda\ell_1 \cos\lambda\ell_{1,1} \sin\lambda\ell_{1,2}$$
$$+ \sin\lambda\ell_1 \sin\lambda\ell_{1,1} \cos\lambda\ell_{1,2}).$$

The functions d corresponding to the simple uncontrolled nodes are

$$d_{1,1}(\lambda) = q_2(\lambda)q_{1,2}(\lambda), \qquad d_{1,2}(\lambda) = q_2(\lambda)q_{1,1}(\lambda), \qquad d_2(\lambda) = q_1(\lambda).$$

Finally,

$$c_k = \max\left(|d_{1,1}(\lambda_k)|, |d_{1,2}(\lambda_k)|, |d_2(\lambda_k)|\right).$$

Now it is easy to determine the situations in which $\mathcal{A}$ is degenerate. If $c_k = 0$
then,

$$|d_{1,1}(\lambda_k)| = |d_{1,2}(\lambda_k)| = |d_2(\lambda_k)| = 0.$$

Taking into account that $\ell_{1,2} = \ell_2$ from the latter equality it follows $\sin \lambda \ell_{1,2} = 0$ and then,

$$\sin \lambda \ell_1 \sin \lambda \ell_{1,1} = 0.$$

If $\sin \lambda \ell_{1,1} = 0$ (resp., $\sin \lambda \ell_1 = 0$) then, necessarily, $\ell_{1,1}/\ell_{1,2}$ (resp. $\ell_1/\ell_{1,2}$) is a rational number. Consequently, we can ensure that $\mathcal{A}$ is non degenerate if the ratios $\ell_{1,1}/\ell_{1,2}$ and $\ell_1/\ell_{1,2}$ are irrational numbers.

Besides, we can give conditions on the lengths that provide lower bounds on the coefficients c_k. If $\alpha \in \mathbb{R}$ is such that $\lambda_k^\alpha c_k \to 0$ then,

$$\lambda_k^\alpha |d_{1,1}(\lambda_k)| \to 0, \quad \lambda_k^\alpha |d_{1,2}(\lambda_k)| \to 0, \quad \lambda_k^\alpha |d_2(\lambda_k)| \to 0.$$

It is easy to see that this implies

$$\lambda_k^{\alpha 2} |\sin \lambda_k \ell_1 \sin \lambda_k \ell_{1,1}| \to 0. \tag{5.101}$$

Then we can apply the results of Appendix A to conclude that (5.101) is impossible if

- the ratios $\ell_{1,1}/\ell_{1,2}, \ell_1/\ell_{1,2}$ and $\ell_1/\ell_{1,1}$ belong to some $\mathbf{B}_\varepsilon$ and $\alpha > 4 + \varepsilon$

or

- the numbers $\ell_1, \ell_{1,1}, \ell_{1,2}$ satisfy the conditions (S) and $\alpha > 2 + \varepsilon$.

Then, in the above cases we will have that there exists a positive constant C such that for every $k \in \mathbb{Z}$,

$$c_k \geq \frac{C}{\lambda_k^\alpha}.$$

Consequently, all the initial states $(\bar{u}_0, \bar{u}_1) \in V^\alpha \times V^{\alpha-1}$ are controllable in a time equal to twice the sum of the lengths of the strings.

6

Some Observability and Controllability Results for General Networks

In this chapter we have gathered some results of general character, which do not impose any restriction on the topological configuration of the networks.

The first one is described in Section 6.1. It concerns the spectral controllability from an exterior node of arbitrary networks, which may, in particular, contain cycles. A condition on the eigenfunctions of the network is given that guarantees its spectral controllability in any time larger than twice its total length.

For tree-shaped networks, that condition coincides with the spectral controllability criterion given in Chapter 5 (Theorem 5.35), except by the fact that there it was possible to obtain information on what happens in the minimal control time as well.

However, for arbitrary networks with more complex structures it is hard to give an algebraic characterization of a condition guaranteeing the spectral controllability. This would require to take into account the specific structure of the graph that supports the network.

In Section 6.2 we present a result of general character, related to the control of an arbitrary network when we allow controls to act on all its nodes. In spite of what may be expected at a first sight, the number of controlled points being so large does not guarantee the exact controllability of the network. It will be shown that in order to reach spectral controllability, it is sufficient to choose only four different control functions, simultaneously applied in several nodes of the network. We shall also determine how each of these four controllers has to be distributed along the nodes of the network.

Finally, Section 6.3 is devoted to show that the Schmidt's theorem stated in Chapter 2 (Theorem 2.7) is sharp in the sense that, if a tree-shaped network has more than one uncontrolled node then it is not exactly controllable in any time $T > 0$.

6.1 Spectral Control of General Networks

6.1.1 Asymptotic Behavior of the Eigenfunctions

The eigenvalues of a network cannot be explicitly computed. It was already impossible to do it for the three string network: in that case, the eigenvalues are determined by the transcendental equation $q(\lambda_k) = 0$, where q is defined by the formula (4.61). However, it is not difficult to obtain certain information on the asymptotic behavior of the sequence of eigenvalues in the general case.

The idea is simple: the eigenvalues of the network may be compared with the eigenvalues of the strings with Dirichlet and Neumann boundary conditions.

To be more precisely, let us denote by $(\mu_n^{i,D}), (\mu_n^{i,N})$ the sequences of eigenvalues of the operator $-\Delta$ on the string $\mathbf{e}_i$ of length ℓ_i with homogeneous boundary conditions of Dirichlet and Neumann type, respectively. Let $(\mu_n^D), (\mu_n^N)$ be the strictly increasing sequences formed by the elements of the sets

$$\bigcup_{i=i}^{M} (\mu_n^{i,D}), \qquad \bigcup_{i=i}^{M} (\mu_n^{i,N}),$$

respectively.

Then, if $(\hat{\mu}_n)$ is the strictly increasing sequence of the eigenvalues of the network, the following holds:

Proposition 6.1. *For every $n \in \mathbb{N}$ the following inequalities are true*

$$\mu_n^N \leq \hat{\mu}_n \leq \mu_n^D.$$

This proposition is proved in [2], [114] for the general case of equations with variable coefficients. For vibrating strings, it seems to have been first stated by Camerer in 1980 (see reference [3] in [114]), though this property has been also studied in [90]. A quite instructive application of these ideas for networks of beams is given in [39].

Let us observe that the eigenvalues $\mu_n^{i,D}$, $\mu_n^{i,N}$ may be computed explicitly:

$$\mu_n^{i,D} = \left(\frac{\pi n}{\ell_i}\right)^2, \qquad \mu_n^{i,N} = \left(\frac{\pi(n-1)}{\ell_i}\right)^2 \qquad n \in \mathbb{N}. \qquad (6.1)$$

Thus, $\mu_n^{i,D} = \mu_{n+1}^{i,N}$ and the same is true for the sequences $(\mu_n^D), (\mu_n^N)$:

$$\mu_n^D = \mu_{n+1}^N.$$

With this, the inequality of Proposition 6.1 becomes

$$\mu_n^N \leq \hat{\mu}_n \leq \mu_{n+1}^N. \qquad (6.2)$$

If we denote $\hat{\lambda}_n := \sqrt{\hat{\mu}_n}$ we get as an immediate consequence of these inequalities the following property of generalized separation of the sequence of eigenvalues

$$\hat{\lambda}_{n+M+1} - \hat{\lambda}_n \geq \sqrt{\mu_{n+M}^N} - \sqrt{\mu_n^N} \geq \pi \min_{i=1,...,M} \left(\frac{1}{\ell_i}\right).$$

Indeed, from the inequalities (6.1) it follows that, for every $i = 1, ..., M$, and every $k \in \mathbb{N}$

$$\sqrt{\mu_{k+1}^{i,N}} - \sqrt{\mu_k^{i,N}} = \frac{\pi}{\ell_i}.$$

On the other hand, for each $n \in \mathbb{N}$, among the $M + 1$ numbers

$$\sqrt{\mu_n^N}, \sqrt{\mu_{n+1}^N}, ..., \sqrt{\mu_{n+M}^N}$$

there are necessarily at least two that correspond to the same value i^* of i. Then

$$\sqrt{\mu_{n+M}^N} - \sqrt{\mu_n^N} \geq \sqrt{\mu_{k+1}^{i^*,N}} - \sqrt{\mu_k^{i^*,N}} = \frac{\pi}{\ell_{i^*}} \geq \pi \min_{i=1,...,M} \left(\frac{1}{\ell_i}\right).$$

In a similar way, as it has been done in Chapter 4 for the three string network, it is possible to obtain asymptotic information on the sequence $(\hat{\mu}_n)$ from the inequalities (6.2). Indeed, if $n(r, (a_n))$ denotes the counting function of the sequence (a_n), that is, $n(r, (a_n))$ is the number of elements of a_n contained on the interval $(0, r)$, then from inequality (6.2) it follows

$$n(r, (\sqrt{\mu_n^N})) - 1 \leq n(r, (\hat{\lambda}_n)) \leq n(r, (\sqrt{\mu_n^N})). \tag{6.3}$$

On the other hand,

$$n(r, (\sqrt{\mu_n^N})) = \sum_{i=1}^{M} n(r, (\sqrt{\mu_n^{i,N}})). \tag{6.4}$$

But $\sqrt{\mu_n^{i,N}} = \pi(n-1)/\ell_i$, and thus

$$n(r, (\sqrt{\mu_n^{i,N}})) = [\frac{r\ell_i}{\pi}] + 1$$

(here, $[\eta]$ denotes the integer part of the real number η). From this inequality we obtain

$$\frac{r\ell_i}{\pi} \leq n(r, (\sqrt{\mu_n^{i,N}})) \leq \frac{r\ell_i}{\pi} + 1$$

and then, from (6.4),

$$\frac{r}{\pi}L \leq n(r, (\sqrt{\mu_n^N})) \leq \frac{r}{\pi}L + M,$$

where L is the sum of the lengths of all the strings entering in the network, the total length of the graph previously denoted by L_A.

Finally, replacing this estimate in (6.3) we obtain

$$\frac{r}{\pi}L - 1 \leq n(r, (\hat{\lambda}_n)) \leq \frac{r}{\pi}L + M. \qquad (6.5)$$

Let us observe that from the inequalities (6.5) it follows that the sequence $(\hat{\lambda}_n)$ has density:

$$D(\hat{\lambda}_n) := \lim_{r \to \infty} \frac{n(r, (\hat{\lambda}_n))}{r} = \frac{L}{\pi}. \qquad (6.6)$$

It is possible to prove that (see, e.g., Problem 1, p. 142 in [116]), for any sequence (a_n),

$$\lim_{r \to \infty} \frac{n(r, (a_n))}{r} = \lim_{n \to \infty} \frac{n}{a_n}.$$

Therefore, from (6.6) we obtain

$$\lim_{n \to \infty} \frac{\hat{\lambda}_n}{n} = \frac{\pi}{L}, \qquad \lim_{n \to \infty} \frac{\hat{\mu}_n}{n^2} = \left(\frac{\pi}{L}\right)^2;$$

that is, asymptotically, the eigenvalues of the network behave as those of one string of length L. This suggests, in view of the fact that for the pointwise control of a string of length ℓ the minimal control time is 2ℓ, that for the control of a network from one of its exterior nodes the minimal control time should be $2L$. In Theorem 6.5 we will prove that this fact is indeed true.

Summarizing the previous results we can formulate

Proposition 6.2. *If $(\hat{\lambda}_n)$ is the strictly increasing sequence formed by the positive square roots of the eigenvalues of the network then,*

1) The counting function of $(\hat{\lambda}_n)$ satisfies

$$\frac{r}{\pi}L - 1 \leq n(r, (\hat{\lambda}_n)) \leq \frac{r}{\pi}L + M.$$

2) The sequence $(\hat{\lambda}_n)$ has upper density

$$D^+(\hat{\lambda}_n) = \frac{L}{\pi}.$$

3) The numbers $\hat{\lambda}_n$ are separated in a generalized sense

$$\hat{\lambda}_{n+M+1} - \hat{\lambda}_n \geq \pi \min_{i=1,\dots,M} \left(\frac{1}{\ell_i}\right).$$

4) For every $T > 2L$ there exist positive numbers γ_n, such that

$$\int_0^T \left| \sum_{n \in \mathbb{Z}} c_n e^{i\lambda_n t} \right|^2 dt \geq \sum_{n \in \mathbb{Z}} \gamma_n^2 |c_n|^2,$$

for every finite sequence (c_n).

$5) \lim_{n \to \infty} \dfrac{\hat{\lambda}_n}{n} = \dfrac{\pi}{L}$

Let us note that the property 4 holds as an immediate consequence of Corollary 3.33 of Theorem 3.32.

Let us recall now a notion from the Theory of Non Harmonic Fourier Series. Let (λ_n) be a sequence of distinct real numbers and denote by Θ the set of all the finite linear combinations

$$f(t) = \sum c_n e^{i\lambda_n t}.$$

The number

$$R(\lambda_n) := \sup \{r : \Theta \text{ is dense } C([-r, r])\}$$

is called *completeness radius* of (λ_n).

The information given by Proposition 6.2 allows us to calculate the completeness radius of the sequence $(\pm\hat{\lambda}_n)$.

Proposition 6.3. *The completeness radius of the sequence* $(\pm\hat{\lambda}_n)$ *is equal to* L.

This assertion is a direct consequence of Theorem 2.3.1 from [50] applied to the sequence $(\pm\hat{\lambda}_n)$. At the same time, that theorem from [50] is a consequence of the famous Beurling-Malliavin theorem allowing to express the completeness radius of a sequence in terms of its density (the details may be found in the original work of A. Beurling and P. Malliavin [21]).

In [50] the following proposition is also proved.

Proposition 6.4 (Haraux and Jaffard, [50]). *Let* (λ_n) *be a sequence of real numbers. Then, the following properties are verified*

1) *For every* $T > 2R(\lambda_n)$ *and every* $k \in \mathbb{Z}$ *there exists a constant* $C_k > 0$ *such that*

$$\int_0^T \left| \sum_{n \in \mathbb{Z}} a_n e^{i\lambda_n t} \right|^2 dt \geq C_k \left| a_k \right|^2, \tag{6.7}$$

for any finite sequence (a_n).

2) *If* $T < 2R(\lambda_n)$ *there is no finite set* $I \subset \mathbb{Z}$ *such that there exists a constant* $C_I > 0$ *with the property that, for some non-trivial set of coefficients* $(\alpha_n)_{n \in I}$, *the inequality*

$$\int_0^T \left| \sum_{n \in \mathbb{Z}} a_n e^{i\lambda_n t} \right|^2 dt \geq C_I \left| \sum_{n \in I} \alpha_n a_n \right|^2, \tag{6.8}$$

is valid for every finite sequence (a_n).

If we apply this result to the sequence $(\pm\hat{\lambda}_n)$, we obtain, in view of Proposition 6.3,

For every $T > 2L$ and $k \in \mathbb{Z}$ there exist positive numbers C_k, such that

$$\int_0^T \left| \sum_{n\in\mathbb{Z}} a_n e^{i\lambda_n t} \right|^2 dt \geq C_k \left| a_k \right|^2,$$

for every finite sequence (a_n).

The drawback of the method of proof of this result, when compared with that used in the proof of property 4) in Proposition 6.2, is that it is not of constructive nature and therefore the coefficients C_k may not be explicitly obtained. However, this result, as we shall see, is a simple consequence of Proposition 6.4 (and the results above on the spectral density), while Proposition 6.2 uses Theorem 3.32 that has been proved more recently.

6.1.2 Application to Control

The results above on the asymptotic behavior of the sequence of eigenvalues allow to obtain the following information in connection to the control of arbitrary networks of strings from one exterior node.

Theorem 6.5. *a) For every $T > 2L$ the following properties of the system (2.11)-(2.16) are equivalent*

1) *the system is approximately controllable in time T;*
2) *the system is spectrally controllable in time T;*
3) *the spectral unique continuation property holds, i. e., $\omega_x^1(\mathbf{v}_1) \neq 0$ is verified by any non-zero eigenfunction $\bar{\omega}$.*

b) When $T < 2L$ system (2.11)-(2.16) is not spectrally controllable; no element of $Z \times Z$ is controllable in time T.

Proof. a) We will prove that 1) $\Rightarrow$ 3) $\Rightarrow$ 2). This, together with the immediate implication 2) $\Rightarrow$ 1) (the spectral controllability is a particular case of the approximate controllability), will give the assertion of the theorem.

1) $\Rightarrow$ 3). Let us observe that if $\varkappa_n = 0$ for some $n = n_0$ then the solution of (2.17)-(2.21)

$$\bar{\phi}(t, x) = \cos \lambda_{n_0} t \; \bar{\theta}_{n_0}(x)$$

satisfies $\phi_x^1(t, \mathbf{v}_1) = 0$ for every $t \in \mathbb{R}$. For this solution $\bar{\phi}$ the unique continuation property from the controlled node is not valid for any value of $T > 0$ and thus, system (2.11)-(2.16) is not approximately controllable in any time $T > 0$. Therefore, 1) $\Rightarrow$ 3).

3) $\Rightarrow$ 2). From Chapter 3 we know that, if the observability inequality

$$\int_0^T \left| \phi_x^1(t, \mathbf{v}_1) \right|^2 dt \geq \sum_{n \in \mathbb{N}} c_n^2 \left(\mu_n \left| \phi_{0,n} \right|^2 + \left| \phi_{1,n} \right|^2 \right), \tag{6.9}$$

is verified for every solution $\bar{\phi}$ of the homogeneous system (2.11)-(2.16) with initial data $(\bar{\phi}_0, \bar{\phi}_1) \in Z \times Z$ then, all the initial data $(\bar{u}_0, \bar{u}_1) \in H \times V'$ satisfying

$$\sum_{n \in \mathbb{N}} \frac{1}{c_n^2} \left| u_{0,n} \right|^2 < \infty, \qquad \sum_{n \in \mathbb{N}} \frac{1}{c_n^2 \mu_n} \left| u_{1,n} \right|^2 < \infty \tag{6.10}$$

are controllable in time T.

Using the formula (2.23) for the solutions of (2.17)-(2.21), the inequality (6.9) may be written as

$$\int_0^T \left| \sum_{n \in \mathbb{N}} \varkappa_n (\phi_{0,n} \cos \lambda_n t + \frac{\phi_{1,n}}{\lambda_n} \sin \lambda_n t) \right|^2 dt \geq \sum_{n \in \mathbb{N}} c_n^2 \left(\mu_n \left| \phi_{0,n} \right|^2 + \left| \phi_{1,n} \right|^2 \right), \tag{6.11}$$

where $(\phi_{0,n})$ and $(\phi_{1,n})$ are finite sequences and $\varkappa_n$ are the values of the normal derivatives of the eigenfunctions at the controlled node:

$$|\varkappa_n| = \theta_{n,x}^1(\mathbf{v}_1).$$

If we denote

$$a_n = \frac{1}{2} \left(\phi_{0,|n|} + \frac{\phi_{1,|n|}}{i\lambda_n} \right),$$

for $n \in \mathbb{Z}_*$, where $\lambda_n = -\lambda_{-n}$ if $n < 0$, we will have

$$\phi_{0,n} = a_n + a_{-n}, \qquad \phi_{1,n} = (a_n - a_{-n})i\lambda_n, \quad n \in \mathbb{N}.$$

With these notations, inequality (6.11) becomes

$$\int_0^T \left| \sum_{n \in \mathbb{Z}_*} a_n \varkappa_{|n|} e^{i\lambda_n t} \right|^2 dt \geq 4 \sum_{n \in \mathbb{N}} c_n^2 \mu_n |a_n|^2, \tag{6.12}$$

for every finite sequence (a_n) of complex numbers with the property $a_{-n} = \overline{a_n}$.

Let us observe now that, as the network is such that no eigenfunction vanishes identically[1] on the controlled string, the eigenvalues μ_n are all simple. Indeed, if $\bar{\psi}$ and $\bar{\varphi}$ are two linearly independent eigenfunctions corresponding to the eigenvalue μ, then the function

$$\bar{\omega} = \varphi_x^1(\mathbf{v}_1)\bar{\psi} - \psi_x^1(\mathbf{v}_1)\bar{\varphi}$$

---

[1] This condition is obviously equivalent to the fact that the normal derivative of the eigenfunction vanishes at the controlled node, since, by definition, the eigenfunctions are equal to zero at the controlled node and satisfy a second order ordinary differential equation.

is also a non-trivial eigenfunction since $\bar{\psi}$ and $\bar{\varphi}$ are linearly independent. Besides

$$\omega_x^1(\mathbf{v}_1) = \varphi_x^1(\mathbf{v}_1)\psi_x^1(\mathbf{v}_1) - \psi_x^1(\mathbf{v}_1)\varphi_x^1(\mathbf{v}_1) = 0,$$

and this contradicts our hypothesis on the network.

Thus, all the eigenvalues being simple, as a result of Proposition 6.2, 4) there exist positive numbers γ_n such that

$$\int_0^T \left| \sum_{n \in \mathbb{Z}_*} a_n \varkappa_n e^{i\lambda_n t} \right|^2 dt \geq 2 \sum_{n \in \mathbb{N}} \gamma_n^2 \varkappa_n^2 |a_n|^2.$$

Therefore, we can conclude that the inequality (6.12) is true with coefficients

$$c_n = \frac{\gamma_n |\varkappa_n|}{\sqrt{2}\lambda_n}.$$

Let us note that all these coefficients are different from zero, since the hypothesis 3) guarantees that $\varkappa_n \neq 0$ for every n. Then, the initial states defined by (6.10) are controllable in time T and, in particular, so are those of the space $Z \times Z$. This means that system (2.11)-(2.16) is spectrally controllable in time T.

b) Let $I \subset \mathbb{N}$ be a finite set. If we apply Corollary 3.15, it follows that the initial state

$$(\bar{u}_0, \bar{u}_1) = (\sum_{n \in I} \alpha_n \bar{\theta}_n, \sum_{n \in I} \beta_n \bar{\theta}_n) \in Z \times Z \tag{6.13}$$

is controllable in time T if, and only if, there exists a constant $C > 0$ such that

$$\int_0^T |\phi_x^1(t, \mathbf{v}_1)|^2 dt \geq C \left(\sum_{n \in I} \alpha_n \phi_{1,n} - \beta_n \phi_{0,n} \right)^2,$$

for every solution $\bar{\phi}$ of (2.17)-(2.21) with initial state $(\bar{\phi}_0, \bar{\phi}_1) \in Z \times Z$.

As a consequence of this, if the initial state $(\bar{u}_0, \bar{u}_1)$ defined by (6.13) is controllable in time T, there exists a constant $C > 0$ such that

$$\int_0^T \left| \sum_{n \in \mathbb{Z}_*} a_n \varkappa_n e^{i\lambda_n t} \right|^2 dt \geq C \left(\sum_{n \in I} \alpha_n (a_n - a_{-n}) i\lambda_n - \beta_n (a_n + a_{-n}) \right)^2$$

$$= C \left(\sum_{n \in I} (\alpha_n i\lambda_n - \beta_n) a_n + (-\alpha_n i\lambda_n - \beta_n) a_{-n} \right)^2$$

$$= C \left(\sum_{n \in I \cup -I} \rho_n a_n \right)^2, \tag{6.14}$$

for every finite sequence (a_n), where

$$\rho_n = \alpha_{|n|} i\lambda_n - \beta_{|n|}.$$

On the other hand, if $T < 2L$ then, since $R(\lambda_n) = L$, we have $T < 2R(\lambda_n)$. Then, in account of Proposition 6.4, 2) we can ensure that there is no sequence satisfying (6.14). Therefore, the initial state $(\bar{u}_0, \bar{u}_1)$ defined by (6.13) is not controllable in time T if $T < 2L$.

Remark 6.6. When $T > 2L$ and the spectral unique continuation property is verified, if we define the space $\mathcal{W}$ as the completion of $Z \times Z$ with the norm

$$|||(\bar{\phi}_0, \bar{\phi}_1)||| := \left\{ \int_0^T \left| \phi_x^1(t, \mathbf{v}_1) \right|^2 dt \right\}^{\frac{1}{2}},$$

then, all the initial states $(\bar{u}_0, \bar{u}_1) \in H \times V'$ such that $(-\bar{u}_1, \bar{u}_0) \in \mathcal{W}'$ (the dual of $\mathcal{W}$) are controllable in time T.

In view of Proposition 6.2, the space $\mathcal{W}'$ contains all those $(-\bar{u}_1, \bar{u}_0)$ that satisfy

$$\sum_{n \in \mathbb{N}} \frac{1}{\gamma_n^2 \varkappa_n^2} \left(|u_{0,n}|^2 + \frac{1}{\mu_n} |u_{1,n}|^2 \right) < \infty,$$

where the coefficients γ_n are computed according to Corollary 3.33 of Theorem 3.32.

Remark 6.7. In general, when $T < 2L$ we do not know whether approximate controllability of (2.11)-(2.16) may hold. Possibly, the available information on the sequences (λ_n) and $(\varkappa_n)$ is not sufficient to conclude.

For the three string network we were able to prove in Section 4.9 of Chapter 4, that the approximate controllability does not hold whenever $T < 2L$. Recall that in that case we constructed explicitly a sequence of solutions for which the unique continuation property fails. The same construction may be done for the star-shaped network with n strings. But extending it to general networks is an open problem.

In [12], the lack of simultaneous approximate controllability of n strings was obtained with the aid of Corollary 3.31. This approach, however, is not appropriate for networks which are not star-shaped, since we do not have sufficient information on the sequence $(\varkappa_n)$.

Finally, unlike the case of tree-shaped networks, the approach in this section does not say whether the spectral controllability still holds in the minimal time $T = 2L$.

6.2 Colored Networks

We consider now a network of N strings controlled at all of it nodes.

The motion of the network may be described by the system

$$\begin{cases} u_{tt}^i - u_{xx}^i = 0 & \text{in } \mathbb{R} \times [0, \ell_i], \ i = 1, ..., N, \\ u^i(t, 0) = v^{k(\mathbf{v}_i^+)}(t) & t \in \mathbb{R}, \qquad i = 1, ..., N-1, \\ u^i(t, \ell_i) = v^{k(\mathbf{v}_i^-)}(t) & t \in \mathbb{R}, \\ u^i(0, x) = u_0^i(x), \ u_t^i(0, x) = u_1^i(x) \ x \in [0, \ell_i], & i = 1, ..., N. \end{cases} \quad (6.15)$$

Here we have denoted by $\mathbf{v}_i^+$, $\mathbf{v}_i^-$ the initial (corresponding to $x = 0$) and final ($x = \ell_i$) nodes of the string $\mathbf{e}_i$, respectively, and $k(\mathbf{v})$ is the index of the node $\mathbf{v}$.

The problem (6.15) is well posed for initial states $(u_0^i, u_1^i) \in L^2(0, \ell_i) \times H^{-1}(0, \ell_i)$, $i = 1, ..., N$, and controls $v^k \in L^2(0, T)$.

This system, being controlled at a large number of points, is expected to have better controllability properties than the control systems studied up to now. As usually, we will say that the initial state $(u_0^i, u_1^i) \in L^2(0, \ell_i) \times H^{-1}(0, \ell_i)$, $i = 1, ..., N$, is controllable in time $T > 0$, if it is possible to choose controls $v^k \in L^2(0, T)$ such that the solution u^i, $i = 1, ..., N$, of (6.15) reaches the rest position in time T:

$$u^i(T, .) = u_t^i(T, .) = 0, \quad i = 1, ..., N.$$

For every $i = 1, ..., M$, we introduce the sets

$$X_i^+ = \left\{ j : \ \mathbf{v}_i^+ \in \mathbf{e}_j \right\}, \qquad X_i^- = \left\{ j : \ \mathbf{v}_i^- \in \mathbf{e}_j \right\},$$

which are, respectively, the sets of the indices of those strings that are incident at the initial and final nodes of the string $\mathbf{e}_i$.

A direct application of HUM guarantees that, if for every $i = 1, ..., M$ there exists a sequence of non-zero real numbers such that

$$\int_0^T \left(|\sum_{j \in X_i^+} \partial_n \phi^j(t, \mathbf{v}_i^+)|^2 + |\sum_{j \in X_i^-} \partial_n \phi^j(t, \mathbf{v}_i^-)|^2 \right) dt$$

$$\geq \sum_{n \in \mathbb{N}} (c_n^i)^2 \left(\mu_n^i (\phi_{0,n}^i)^2 + (\phi_{1,n}^i)^2 \right), \qquad (6.16)$$

for every solution $\bar{\phi} = (\phi^1, ..., \phi^N)$ of the homogeneous problem (6.15) with $v^{k(\mathbf{v}^\pm)}$, for all node $\mathbf{v}$, then, the initial states (u_0^i, u_1^i), $i = 1, ..., N$, verifying

$$\sum_{n \in \mathbb{N}} \frac{(u_{0,n}^i)^2}{(c_n^i)^2} < \infty, \qquad \sum_{n \in \mathbb{N}} \frac{(u_{1,n}^i)^2}{\mu_n^i (c_n^i)^2} < \infty,$$

are controllable in time T.

Let us remark that the homogeneous problem is a set of N uncoupled wave equations with homogeneous Dirichlet boundary conditions. The coupling in the original controlled system (6.15) relies on the fact that the same control acts on all the string components that meet at each node. At the level of observability, in (6.16), this corresponds to the fact that the "observed quantity" in every node is the sum of the normal derivatives of the solutions corresponding to those strings that are coupled it that node. Let us observe that this is a local problem, in the sense that in the observability inequality for every string $\mathbf{e}_i$ only those solutions corresponding to strings that have common nodes with $\mathbf{e}_i$ are present.

Recall that for the simultaneous control of n strings we have proved that if T_i^+ and T_i^- are the sums of the lengths of all the strings that are incident to $\mathbf{v}_i^+$ and $\mathbf{v}_i^-$, respectively, then the following inequalities are verified

$$\int_0^{2T_i^+} |\sum_{j \in X_i^+} \partial_n \phi^j(t, \mathbf{v}_i^+)|^2 dt \geq \sum_{n \in \mathbb{N}} (c_n^i)^2 \left(\mu_n^i (\phi_{0,n}^i)^2 + (\phi_{1,n}^i)^2 \right), \quad (6.17)$$

$$\int_0^{2T_i^-} |\sum_{j \in X_i^-} \partial_n \phi^j(t, \mathbf{v}_i^-)|^2 dt \geq \sum_{n \in \mathbb{N}} (c_n^i)^2 \left(\mu_n^i (\phi_{0,n}^i)^2 + (\phi_{1,n}^i)^2 \right), \quad (6.18)$$

with coefficients[2] that may be explicitly computed whenever ℓ_p/ℓ_q are irrational numbers for every $p, q \in X_i^+$ for (6.17) and $p, q \in X_i^-$ for (6.18).

Consequently, we can indicate conditions on the lengths of the strings, precisely those given for the simultaneous control of n strings, guaranteeing the controllability of system (6.15) in explicitly characterized spaces. In particular, under the irrationality hypotheses mentioned above, the system is spectrally controllable in any time

$$T^* \geq 2 \max_{i=1,\ldots,N} \left\{ T_i^+, T_i^- \right\}.$$

Now we attempt to reduce the number of different functions used to control system (6.15) by applying the same control function at several nodes. We introduce a partition of the set $\{1, 2, \ldots, N\}$ of indexes of the nodes of the network:

$$\{1, 2, \ldots, N\} = K_1 \cup \cdots \cup K_r.$$

This partition is such that there is no string having its two nodes in the same set K_k. Two nodes are said to be equivalent if their indexes belong to the same class. A simple way of building this partition is to paint the nodes using r different colors such that no string has both nodes of the same color. With this, equivalent nodes are those of the same color. This equivalence relation is denoted by $\sim$.

In Figure 6.1 we have represented a network, whose nodes have been painted with four colors, represented by the symbols: $\blacklozenge, \blacktriangle, \blacktriangledown, \bullet$. For this network, the smallest number of colors that allows painting the nodes, fulfilling the previous requirements, is four.

---

[2] The coefficients c_n^i in the inequalities (6.17) and (6.18) are different. We have denoted them with the same symbols to avoid to make the notations even more complex.

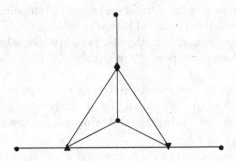

Fig. 6.1. A network with colored nodes

Now we add an additional restriction to system (6.15): $v_p = v_q$ if the nodes $\mathbf{v}_p$ and $\mathbf{v}_q$ are of the same color. In other words, the same control is applied to all nodes in the same equivalence class.

It is easy to see that this restriction leads to the same observability inequality as before, except by the fact that now the sets X_i^+ and X_i^- should be replaced by

$$X_i^+ = \bigcup_{\mathbf{v} \sim \mathbf{v}_i^+} \{j: \quad \mathbf{v} \in \mathbf{e}_j\}, \qquad X_i^- = \bigcup_{\mathbf{v} \sim \mathbf{v}_i^-} \{j: \quad \mathbf{v} \in \mathbf{e}_j\},$$

which are the sets of the indices of the strings, which have some node of the same color as the initial and final nodes of $\mathbf{e}_i$, respectively.

We may conclude that, if the lengths of the strings satisfy $\ell_p/\ell_q \notin \mathbb{Q}$ for all the indices $p \neq q$ such that some of the nodes of the string $\mathbf{e}_p$ are of the same color as one of the nodes of $\mathbf{e}_q$ then system (6.15) is spectrally controllable in any time

$$T \geq 2 \max_{k=1,\ldots,r} (L_k),$$

where L_k is the sum of the lengths of all strings having some node of color k.

To avoid complex notations and simplify the presentation, we will replace the previous conditions by the following, clearly more restrictive ones:

1) the ratios ℓ_p/ℓ_q are irrational numbers for all the indices $p \neq q$;
2) $T \geq 2L$, L being the sum of the lengths of all the strings of the network.

Under these hypotheses, the minimal number of different control functions needed to reach the spectral controllability of the network is equal to the minimal number of colors which are necessary to paint the vertices of the graph such that no edge has its vertices of the same color. This is the classical problem on colored graphs (and this is equivalent to painting a map).The solution, the famous Four Colors Theorem, [8], asserts that if the graph is planar, four colors are sufficient. [3]

Let us observe now that we have actually obtained two inequalities for every string: (6.17) and (6.18), while only one of them suffices to prove the corresponding observability inequality. That is why we may assume that the control associated with one of the colors vanishes (this is, indeed a particular choice of the control). Note however that imposing to the nodes in that class to have zero displacement is indeed introducing a control, the zero one.

Summarizing the previous results, the following has been proved:

Proposition 6.8. *If the network is supported on a planar graph and the lengths of its strings and T satisfy the conditions 1) and 2) above then, four different controls are sufficient for the system (6.15) to be spectrally controllable in time T. Besides, one of those functions may be chosen to be identically equal to zero.*

Remark 6.9. The condition requiring that the extremes of the strings are of distinct colors is natural if one expects at least the approximate controllability property of the system to hold. This is so because it is impossible to control a string using the same control function in both of its extremes. Indeed, the observability inequality associated to that problem for a string occupying the space interval $(0, \ell)$ would be

$$\int_0^T |\phi_x(t,0) - \phi_x(t,\ell)|^2 dt \geq \sum_{n \in \mathbb{N}} c_n^2 \left(\mu_n \phi_{0,n}^2 + \phi_{1,n}^2 \right),$$

where ϕ is the solution of the wave equation $\phi_{tt} - \phi_{xx} = 0$ with boundary conditions $\phi(t,0) = \phi(t,\ell) = 0$. It suffices to take

$$\phi(t,x) = \cos \frac{2\pi}{\ell} t \, \sin \frac{2\pi}{\ell} x,$$

to see that this inequality cannot be true. In this example, in fact, $\phi_x(t,0) - \phi_x(t,\ell) = 0$, and thus, the approximate controllability does not hold either.

6.3 Optimality of Theorem 3.2.7

In this section we will prove that Theorem 2.7 in Chapter 3 is sharp in the sense that, if in a tree-shaped network there are at least two uncontrolled nodes, then there exist initial states $(\bar{u}_0, \bar{u}_1) \in H \times V'$ of the network that are not controllable in any finite time T.

---

[3] This is an intuitive result, but a "purely mathematical" rigorous proof is not known. The proof in [8], which is the one available so far, requires the aid of computers.

The proof is based on the fact that if there are two uncontrolled nodes, then there is a simple path formed by consecutive strings connecting those nodes. If there exists $T > 0$ such that every initial state $(\bar{u}_0, \bar{u}_1) \in H \times V'$ is controllable in time T then, the system of serially connected strings with controls at the coupling points would be exactly controllable as well. In the following Subsection 6.3.1. We concentrate on studying in detail that latter system. We will prove that, actually, it is never exactly controllable, independently of the value of T or the lengths of the strings. From this fact it follows that a network with more than two uncontrolled nodes is never exactly controllable.

6.3.1 Simultaneous Control of Serially Connected Strings

Let us suppose that we have N strings of lengths $\ell_1, ..., \ell_N$, which are serially connected and that in every coupling point a control acts to determine the displacement of that point.

The motion of these strings is described by the system of equations

$$\begin{cases} u_{tt}^k - u_{xx}^k = 0 & \text{in } \mathbb{R} \times [0, \ell_k], \ k = 1, ..., N, \\ u^k(t, \ell_k) = u^{k+1}(t, 0) = v^k(t) & t \in \mathbb{R}, \qquad k = 1, ..., N-1, \\ u^1(t, 0) = u^N(t, \ell_N) = 0 & t \in \mathbb{R}, \\ u^k(0, x) = u_0^k(x), \ u_t^k(0, x) = u_1^k(x) & x \in [0, \ell_k], \qquad k = 1, ..., N. \end{cases}$$

$$(6.19)$$

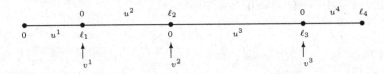

Fig. 6.2. Four serially connected strings with controls v^1, v^2, v^3 at the connection points.

For $T > 0$, this problem is well posed for initial states $(u_0^k, u_1^k) \in L^2(0, \ell_k) \times H^{-1}(0, \ell_k)$, $k = 1, ..., N$, and controls $v^k \in L^2(0, T)$. The corresponding homogeneous problem is also well posed for $(u_0^k, u_1^k) \in H_0^1(0, \ell_k) \times L^2(0, \ell_k)$.

Let us note that this problem is a particular case of that on colored networks studied in Section 6.2. Thus, we can give conditions on the lengths of the strings guaranteeing that the system is spectrally controllable. However, our aim is now to prove the existence of initial data $(u_0^k, u_1^k) \in L^2(0, \ell_k) \times H^{-1}(0, \ell_k)$, $k = 1, ..., N$, which are not controllable in any finite time $T > 0$, independently of the values of the lengths of the strings.

Once again, applying HUM, it follows that the system (6.19) is exactly controllable in time T if, and only if, there exists a constant $C > 0$ such that the solutions of the homogeneous system $\bar{\phi} = (\phi^1, ..., \phi^N)$, which in this case corresponds to N wave equations with homogeneous Dirichlet boundary conditions, verify

$$\sum_{k=1}^{N-1} \int_0^T \left| \phi_x^k(t, \ell_k) - \phi_x^{k+1}(t, 0) \right|^2 dt \geq C \sum_{k=1}^N \mathbf{E}_k, \qquad (6.20)$$

for $k = 1, ..., N$, where $\mathbf{E}_k$ is the energy of the solution ϕ^k, a conserved quantity.

Let us note that the inequality (6.20) cannot be true for arbitrary values of the lengths of the strings. Indeed, if, for example, all the lengths coincide and are equal to ℓ then the functions

$$\phi^k(t, x) = (-1)^k \sin \frac{t\pi}{\ell} \sin \frac{x\pi}{\ell}, \qquad k = 1, ..., N,$$

are solutions of the homogeneous system (6.19) and besides

$$\phi_x^k(t, \ell_k) - \phi_x^{k+1}(t, 0) = (-1)^k \frac{\pi}{\ell} \left(\sin \frac{t\pi}{\ell} \cos \frac{x\pi}{\ell} \big|_{x=\ell} + \sin \frac{t\pi}{\ell} \cos \frac{x\pi}{\ell} \big|_{x=0} \right) = 0,$$

for $k = 1, ..., N$. But the energy of these solutions does not vanish, since the solutions are non-trivial. Then, inequality (6.20) is not true. Moreover, the corresponding unique continuation property being false, the system is not even approximately controllable.

Similar examples may be easily given whenever the lengths of the strings satisfy the conditions $\ell_{k+1}/\ell_k \in \mathbb{Q}$ for every k. On the other hand, if the ratio ℓ_{k+1}/ℓ_k is an irrational number for some k, then the unique continuation property holds and then system (6.19) is approximately controllable.

However, as we have pointed out above, the inequality (6.20) is never valid, independently of the values of the lengths of the strings. Our aim is to prove this assertion.

For every $k = 1, ..., N$, the solution ϕ^k may be expressed as

$$\phi^k(t, x) = \sum_{n \in \mathbb{N}} \left(\phi_{0,n}^k \cos \lambda_n^k t + \frac{\phi_{1,n}^k}{\ell_k} \sin \lambda_n^k t \right) \sin \lambda_n^k x,$$

where $\lambda_n^k = n\pi/\ell_k$ are the eigenvalues of the k-th string and $(\phi_{0,n}^k)$, $(\phi_{1,n}^k)$ are the sequences of Fourier coefficients of the initial data ϕ_0^k, ϕ_1^k in the basis $(\sin \lambda_n^k x)_{n \in \mathbb{N}}$ of $L^2(0, \ell_k)$.

Then

$$\phi^k(t, x) = \sum_{n \in \mathbb{Z}^*} a_n^k e^{i\sigma_n^k t} \sin \lambda_n^k x,$$

where

$$\sigma_n^k = \operatorname{sgn}(n)\lambda_n^k, \qquad a_n^k = \frac{1}{2}\left(\phi_{0,|n|}^k + \frac{\phi_{1,|n|}^k}{i\lambda_n^k}\right).$$

The inequality (6.20) can now be written as

$$\sum_{k=1}^{N-1} \int_0^T \left| \sum_{n\in\mathbb{Z}^*} \left((-1)^n \lambda_n^k a_n^k e^{i\sigma_n^k t} - \lambda_n^{k+1} a_n^{k+1} e^{i\sigma_n^{k+1} t}\right)\right|^2 dt \qquad (6.21)$$

$$\geq C \sum_{k=1}^{N} \sum_{n\in\mathbb{Z}^*} \left|\lambda_n^k a_n^k\right|^2. \qquad (6.22)$$

Our aim is to construct sequences (a_n^k), $k = 1, ..., N$ for which the inequality (6.22) is not verified.

In order to simplify the notations, we assume $N = 2$. Let (σ_n) be the increasing sequence formed by the elements of the sequences (σ_n^1) and (σ_n^2). Define (α_n) by

$$\alpha_n = (-1)^m \lambda_m^1 a_m^1, \qquad \text{if } \sigma_n = \sigma_m^1,$$
$$\alpha_n = -\lambda_m^2 a_m^2 \qquad \text{if } \sigma_n = \sigma_m^2.$$

With this, the inequality (6.22) becomes

$$\int_0^T \left| \sum_{n\in\mathbb{Z}^*} \alpha_n e^{i\sigma_n t} \right|^2 dt \geq C \sum_{n\in\mathbb{Z}^*} |\alpha_n|^2. \qquad (6.23)$$

Note that the latter inequality could be obtained as a consequence of the classical Ingham inequality, if there would be some uniform gap between the numbers σ_n. We will see that this is not the case.

As $\sigma_{m+1}^1 - \sigma_m^1 = \pi/\ell_1$, $\sigma_{m+1}^2 - \sigma_m^2 = \pi/\ell_2$ we can ensure that $\sigma_{n+2} - \sigma_n \geq \pi \min\{1/\ell_1, 1/\ell_2\}$; however, we claim that $\liminf_{n\to\infty}(\sigma_{n+1} - \sigma_n) = 0$. Indeed, it suffices to note that the number ℓ_1/ℓ_2 may be approximated by rational ones, that is, there exist sequences (p_k), (q_k) of integer numbers such that

$$\lim_{k\to\infty}\left(\frac{\ell_1}{\ell_2} - \frac{p_k}{q_k}\right) = 0.$$

This is equivalent to $\sigma_{p_k}^1 - \sigma_{q_k}^2 \to 0$. Thus, the elements of the sequence (σ_n) may get close.

This lack of uniform gap between the numbers σ_k not only makes impossible to apply the Ingham inequality, but also that (6.23) is not true. The following holds:

Proposition 6.10. *There is no positive constant C such that the inequality (6.20) is verified by all the solutions of the homogeneous system (6.19) with initial states $(\phi_0^k, \phi_1^k) \in H_0^1(0, \ell_k) \times L^2(0, \ell_k)$, $k = 1, ..., N$.*

Proof. The key element of the proof is the Dirichlet theorem of simultaneous approximation of real numbers by rational ones (for more details see [26], Section I.5):

If $\xi^1, ..., \xi^M$ are real numbers then, for every $\varepsilon > 0$ and an infinite number of values of $p \in \mathbb{Z}$ there exist integer numbers $q_i(p)$, $i = 1, ..., M$ such that

$$|p\xi_i - q_i(p)| \le \varepsilon \qquad i = 1, ..., M.$$

Let us fix $\varepsilon > 0$ and choose $\xi^i = \ell_{i+1}/\ell_1$, $i = 1, .., N-1$. Applying the Dirichlet theorem to the numbers ξ^i, $i = 1, .., N-1$ it holds that there exist infinite values of p for which the following inequality is verified

$$\left| \frac{p\pi}{\ell_1} - \frac{q_i(p)\pi}{\ell_{i+1}} \right| \le \varepsilon_1 = \varepsilon \max_{i=1,..,N-1} \left(\frac{1}{\ell_{i+1}} \right),$$

and that is

$$\left| \lambda_p^1 - \lambda_{q_i(p)}^{i+1} \right| \le \varepsilon_1. \qquad (6.24)$$

Now denote by (σ_n) the increasing sequence formed by the positive square roots of the eigenvalues λ_n^k of all the strings.

For each value p whose existence was ensured by the Dirichlet theorem, let us define $m(p)$ by

$$\sigma_{m(p)} = \min \left\{ \lambda_p^1, \lambda_{q_1(p)}^2, ..., \lambda_{q_{N-1}(p)}^i \right\}.$$

Then, for an infinite number of values of $p \in \mathbb{Z}$ the following inequality is true:

$$\left| \sigma_{m(p)+N-1} - \sigma_{m(p)} \right| \le \varepsilon_1.$$

Since the elements $\sigma_{m(p)}, \sigma_{m(p)+1}, ..., \sigma_{m(p)+N-1}$ are close we can ensure that, among them, there is exactly one of the eigenvalues of every string. Let $n_k(p)$ be such that

$$\lambda_{n_k(p)}^k \in \left\{ \sigma_{m(p)}, \sigma_{m(p)+1}, ..., \sigma_{m(p)+N-1} \right\}$$

(this value is unique).

Then it will hold

$$\left| \lambda_{n_k(p)}^k - \lambda_{n_{k'}(p)}^{k'} \right| \le \varepsilon_1, \qquad (6.25)$$

for all $k, k' = 1, ..., N$.

Let us consider now, for every $k = 1, ..., N-1$, the following solutions of the homogeneous version of (6.19):

$$\phi_p^k(t, x) = \frac{1}{2\lambda_{n_k(p)}^k} \cos 2\lambda_{n_k(p)}^k t \, \sin 2\lambda_{n_k(p)}^k x,$$

whose energy is $\mathbf{E}_k = \ell_k/2$.

On the other hand,

$$\phi_{p,x}^{k}(t,\ell_k) - \phi_{p,x}^{k+1}(t,0) = \cos 2\lambda_{n_k(p)}^{k}t - \cos 2\lambda_{n_{k+1}(p)}^{k+1}t.$$

Then, from this inequality

$$\int_0^T \left|\phi_{p,x}^{k}(t,\ell_k) - \phi_{p,x}^{k+1}(t,0)\right|^2 dt \leq \left|\lambda_{n_k(p)}^{k} - \lambda_{n_{k+1}(p)}^{k+1}\right|^2 \frac{T^3}{3}.$$

Here we have used the inequality

$$\int_0^T \left(\cos xt - \cos yt\right)^2 \leq \frac{T^3}{3}|x-y|^2,$$

which is easily proved with the help of the mean value theorem.

In account of (6.25), we may conclude that

$$\int_0^T \left|\phi_{p,x}^{k}(t,\ell_k) - \phi_{p,x}^{k+1}(t,0)\right|^2 dt \leq C \left|\lambda_{n_k(p)}^{k} - \lambda_{n_{k+1}(p)}^{k+1}\right|^2 \leq \frac{T^3}{3}\varepsilon_1^2.$$

Finally, if the inequality (6.20) were true we would obtain

$$\cdot \frac{C}{2}\sum_{k=1}^{N}\ell_k = C\sum_{k=1}^{N}\mathbf{E}_k \leq \frac{T^3}{3}\varepsilon_1^2 = \varepsilon^2 \max_{i=1,..,N-1}\left(\frac{1}{\ell_{i+1}}\right)^2 \frac{T^3}{3},$$

what is impossible, since ε may be chosen arbitrarily small.

Remark 6.11. The problem of controlling N strings connected in a cycle with controls in all the nodes may be studied exactly in the same way. This problem is also described by the system (6.19) where the conditions $u^1(t,0) = u^N(t,\ell_N) = 0$ are replaced by $u^1(t,0) = u^N(t,\ell_N) = v_N(t)$. In Chapter VII in [9] the reader may find a proof of the lack of exact controllability in this case, based on the method of moments.

Simultaneous Observation and Control from an Interior Region

This chapter is devoted to the simultaneous control of strings with different densities from a common interior region.

This study is mainly motivated by the following fact. If we perform the changes of variables $x \to \ell_1 x$, $x \to \ell_2 x$ in the equations of system (4.10) for the simultaneous control of strings with density equal to one, from one of the exterior nodes, we obtain

$$
\begin{cases}
\ell_k^2 u_{tt}^1 - u_{xx}^k = 0 & \text{in } \mathbb{R} \times [0,1], \quad k = 1, 2, \\
u^k(.,0) = v, \quad u^k(.,1) = 0 & \text{in } \mathbb{R}, \\
u^k(0,.) = u_0^k, \quad u_t^k(0,.) = u_1^k & \text{in } [0,1].
\end{cases}
\tag{7.1}
$$

Thus, the simultaneous control of two strings of lengths ℓ_1 and ℓ_2 from one of the exterior nodes may be also viewed as the simultaneous control from one end of two strings of equal lengths with different densities ℓ_1^2 and ℓ_2^2.

As we have seen in Chapter 4, the answer to this problem depends on the degree of irrationality of the number ℓ_1/ℓ_2. More precisely, ℓ_1/ℓ_2 needs to be irrational to guarantee that all the Fourier components of the solutions are observable, but, even when ℓ_1/ℓ_2 is irrational, the space of data in which the controllability holds also depends on the value of that ratio.

All this suggests to study the same problem when the control acts over an interior region of the strings, a situation where one expects the control mechanism to be much more robust and the control result not to depend on the ratio ℓ_1/ℓ_2. We refer to [42] for the analysis of the connection (through a singular limit) of these problems of pointwise control and control on a subinterval for one single string.

This issue is analyzed in this chapter. We first consider the one-dimensional case to later address the multi-dimensional one. In one space dimension we give a complete answer to the problem and confirm that this control mechanism, i. e. distributing the control along a subinterval, is much more robust than simply controlling on the boundary. In the multi-dimensional one we simply address the case where the control is applied everywhere in the domain.

7.1 Simultaneous Interior Control of Two Strings

7.1.1 Statement of the Problem

Let ℓ_1, ℓ_2 be positive numbers and ω an interval contained in $(0, \ell_1) \cap (0, \ell_2)$.
    Let us consider the system

$$
\begin{cases}
\rho_k^2 u_{tt}^k - u_{xx}^k + f\chi_\omega = 0 & \text{in } \mathbb{R} \times [0, \ell_k], \\
u^k(.,0) = u^k(., \ell_k) = 0 & \text{in } \mathbb{R}, \\
u^k(0,.) = u_0^k, \quad u_t^k(0,.) = u_1^k & \text{in } [0, \ell_k], \quad k = 1, 2,
\end{cases}
\tag{7.2}
$$

where $f \in L^2_{\text{loc}}(\mathbb{R}^2)$ and χ_ω is the characteristic function of the interval ω.
    This system describes the motion of two strings e_1 and e_2 of lengths ℓ_1, ℓ_2 and densities ρ_1^2, ρ_2^2, respectively, which are simultaneously controlled by means of the same force f localized on the interval ω.
    System (7.2) is well posed for initial states

$$
(u_0^k, u_1^k) \in \mathcal{W}_k := H_0^1(0, \ell_k) \times L^2(0, \ell_k), \qquad k = 1, 2,
$$

with a control force $f \in L^1(0, T; L^2(\omega))$. More precisely, under the previous assumptions on the initial data (u_0^k, u_1^k) and the control force f, system (7.2) admits a unique solution in the energy space

$$
(u^k, u_t^k) \in C([0, T]; \mathcal{W}_k), \qquad k = 1, 2.
$$

    We study the following control problem for (7.2): *given $T > 0$, to determine for which initial states (u_0^k, u_1^k), $k = 1, 2$, the function f may be chosen such that*

$$
u^k(T, .) = u_t^k(T, .) = 0, \quad k = 1, 2.
$$

We will say that system (7.2) is *exactly controllable in time T* if all the initial states $(u_0^k, u_1^k) \in \mathcal{W}_k$, $k = 1, 2$, are controllable in time T.
    The application of the HUM guarantees that (7.2) is exactly controllable in time T if, and only if, there exists a constant $C > 0$ such that

$$
C \int_\omega \int_0^T |\phi^1(t, x) + \phi^2(t, x)|^2 dt dx \geq ||(\phi_0^1, \phi_1^1)||_{L^2 \times H^{-1}}^2 + ||(\phi_0^2, \phi_1^2)||_{L^2 \times H^{-1}}^2,
\tag{7.3}
$$

for all the solutions ϕ^1, ϕ^2 of the homogeneous equations

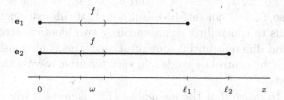

Fig. 7.1. Two strings e_1 and e_2 of different densities controlled simultaneously from the common interval ω.

$$\begin{cases} \rho_k^2 \phi_{tt}^k - \phi_{xx}^k = 0 & \text{in } \mathbb{R} \times [0, \ell_k], \quad k = 1, 2, \\ \phi^k(.,0) = \phi^k(.,\ell_k) = 0 & \text{in } \mathbb{R}, \\ \phi^k(0,.) = \phi_0^k, \quad \phi_t^k(0,.) = \phi_1^k & \text{in } [0, \ell_k]. \end{cases} \tag{7.4}$$

The solutions of (7.4) are given by the formula

$$\phi^k(t,x) = \sum_{n \in \mathbb{N}} \left(\phi_{0,n}^k \cos \frac{n\pi}{\rho_k \ell_k} t + \frac{\rho_k \ell_k}{n\pi} \phi_{1,n}^k \sin \frac{n\pi}{\rho_k \ell_k} t \right) \sin \frac{n\pi}{\ell_k} x, \tag{7.5}$$

where $(\phi_{0,n}^k)$, $(\phi_{1,n}^k)$ are the sequences of Fourier coefficients of ϕ_0^k, ϕ_1^k, respectively, in the basis $(\sin(n\pi x/\ell_k))_{n \in \mathbb{N}}$ of $L^2(0, \ell_k)$.

If we denote

$$a_n^k = \frac{1}{2} \left(\phi_{0,|n|}^k + \frac{i\rho_k \ell_k}{n\pi} \phi_{1,|n|}^k \right), \quad k = 1, 2, \quad n \in \mathbb{Z}_*,$$

the formula (7.5) may be rewritten as

$$\phi^k(t,x) = \sum_{n \in \mathbb{Z}_*} a_n^k e^{\frac{in\pi}{\rho_k \ell_k} t} \sin \frac{n\pi}{\ell_k} x.$$

Note that, by the definition of (a_k^n) we have $\overline{a_k^n} = a_{-k}^n$.

With these notations, the inequality (7.3) is equivalent to

$$C \int_\omega \int_0^T \left| \sum_{n \in \mathbb{Z}_*} a_n^1 e^{\frac{in\pi}{\rho_1 \ell_1} t} \sin \frac{n\pi}{\ell_1} x + a_n^2 e^{\frac{in\pi}{\rho_2 \ell_2} t} \sin \frac{n\pi}{\ell_2} x \right|^2 dt\, dx$$

$$\geq \sum_{n \in \mathbb{N}} \left(|a_n^1|^2 + |a_n^2|^2 \right), \tag{7.6}$$

for all finite complex sequences $(a_n^1)_{n \in \mathbb{Z}_*}$, $(a_n^2)_{n \in \mathbb{Z}_*}$ satisfying $a_{-n}^1 = \overline{a_n^1}$, $a_{-n}^2 = \overline{a_n^2}$.

Obviously, inequality (7.6) is impossible if $\ell_1 = \ell_2$ and $\rho_1 = \rho_2$. Indeed, it suffices to take, e. g., $a_1^1 = -a_1^2 \neq 0$ and $a_n^1 = -a_n^2 = 0$ for $n \neq \pm 1$, to see that, in this case, (7.6) is not satisfied. But note that this extremely degenerate case corresponds to controlling simultaneously two identical strings with the same control and different initial configurations. This is obviously impossible in general, since the control depends in a very sensitive way on the initial data to be controlled.

Our aim is to prove that the inequality (7.6) is verified whenever $\rho_1 \neq \rho_2$ if T is sufficiently large. This is the object of the next subsection.

7.1.2 Control of Strings with Different Densities

Now we consider the case when the densities of the strings are different, i.e., $\rho_1 \neq \rho_2$. The following holds

Theorem 7.1. *If $\rho_1 \neq \rho_2$ and $T > T_0 := 2\max(\rho_1\ell_1, \rho_2\ell_2)$ then the inequality (7.6) is verified for all the finite complex sequences $(a_n^1)_{n\in\mathbb{Z}_*}$, $(a_n^2)_{n\in\mathbb{Z}_*}$ satisfying $a_{-n}^1 = \overline{a_n^1}$, $a_{-n}^2 = \overline{a_n^2}$.*

Corollary 7.2. *The strings e_1 and e_2 are simultaneously exactly controllable in time $T > T_0$ if $\rho_1 \neq \rho_2$.*

Remark 7.3. This result shows an important difference between the control from an extreme of the strings and the control from an arbitrarily small interior region. Recall that, according to Corollary 4.7, all the initial states from $\left(H_0^1(0,1) \times L^2(0,1)\right)^2$ for the system (7.1) are controllable in time $T \geq 2(\ell_1 + \ell_2)$ if, and only if, the ratio ℓ_1/ℓ_2 belongs to the set[1] $\mathcal{F}$, which has zero Lebesgue measure. Besides, the exact controllability property of (7.1) in the optimal space $\left(L^2(0,1) \times H^{-1}(0,1)\right)^2$ is never reached, independently of the values of ℓ_1 and ℓ_2, and there is at least a loss of one derivative in the space of controllable data. Controlling from an interior subinterval provides a much more robust control mechanism, since we may recover the controllability in the sharp space and without irrationality assumptions on the parameters entering in the equations.

In order to prove Theorem 7.1 we will use the scheme of proof of Theorem 4.3 relative to the simultaneous control of two strings from one extreme. The idea is quite simple: given an interval $\omega \subset \mathbb{R}$, we construct another interval $\omega' \subset \omega$ and $T' < T$, and a continuous operator

$$\mathbf{B} : L^2((0,T) \times \omega) \to L^2((0,T') \times \omega')$$

such that, if ϕ^1, ϕ^2 are solutions of (7.4) then $\mathbf{B}\phi^1 = 0$ and, besides, there exists a constant $C > 0$ such that

---

[1] Recall that $\mathcal{F}$ is constituted by the real numbers having a development in continuous fraction $[a_0, a_1, ..., a_n, ...]$ with bounded (a_n).

$$C \int_{\omega'} \int_0^{T'} |\mathbf{B}\phi^2|^2 dtdx \geq ||(\phi_0^2, \phi_1^2)||_{L^2 \times H^{-1}}^2.$$

Then we will have

$$C \int_{\omega} \int_0^T |\phi^1 + \phi^2|^2 dtdx \geq C \int_{\omega'} \int_0^{T'} |\mathbf{B}\phi^2|^2 dtdx \geq ||(\phi_0^2, \phi_1^2)||_{L^2 \times H^{-1}}^2.$$

The inequality

$$C \int_{\omega} \int_0^T |\phi^1 + \phi^2|^2 dtdx \geq ||(\phi_0^1, \phi_1^1)||_{L^2 \times H^{-1}}^2$$

may be obtained in an analogous way.

Let us fix $\omega = (\omega_1, \omega_2) \subset \mathbb{R}$ and define, for $a > 0$, the linear operator $\mathbf{B}_a$ that acts over a function $\phi(t, x)$ according to the formula

$$\begin{aligned}
\mathbf{B}_a \phi(t, x) := & \phi(t + 2(x - \omega_1), x + a(x - \omega_1)) \\
& -\phi(t + (x - \omega_1), x + 2a(x - \omega_1)) \\
& -\phi(t + (x - \omega_1), x) + \phi(t, x + a(x - \omega_1)).
\end{aligned}$$

Let us observe that, since $\omega_1 < \omega_2$ and $T > 0$, it is possible to choose for every $a > 0$ a number $\hat{\omega}_2 \in (\omega_1, \omega_2)$ such that

$$\hat{\omega}_2 < \frac{\omega_2 + 2a\omega_1}{1 + 2a} \quad \text{and} \quad \hat{T} := T - 2(\hat{\omega}_2 - \omega_1) > 0. \tag{7.7}$$

Proposition 7.4. *If $\hat{\omega}_2$ and $\hat{T}$ satisfy (7.7) then the operator $\mathbf{B}_a$ is continuous from $L^2((0, T) \times (\omega_1, \omega_2))$ to $L^2((0, \hat{T}) \times (\omega_1, \hat{\omega}_2))$, that is, there exists a constant $C > 0$ such that*

$$C \int_{\omega_1}^{\omega_2} \int_0^T |\phi(t, x)|^2 dtdx \geq \int_{\omega_1}^{\hat{\omega}_2} \int_0^{\hat{T}} |\mathbf{B}_a \phi(t, x)|^2 dtdx,$$

for every function ϕ for which both integrals are defined.

Proof. Let us observe that

$$\mathbf{B}_a \phi(t, x) = \sum_{(p,q) \in S} (-1)^p \phi(t + p(x - \omega_1), x + qa(x - \omega_1)),$$

where

$$S = \{(2, 1), (1, 2), (0, 1), (1, 0)\}.$$

Then,

$$\int_{\omega_1}^{\hat{\omega}_2} \int_0^{\hat{T}} |\mathbf{B}_a \phi(t, x)|^2 dtdx$$

$$\leq 4 \sum_{(p,q) \in S} \int_{\omega_1}^{\hat{\omega}_2} \int_0^{\hat{T}} |\phi(t + pa(x - \omega_1), x + qa(x - \omega_1))|^2 dtdx. \tag{7.8}$$

To estimate the integrals

$$\int_{\omega_1}^{\hat{\omega}_2} \int_0^{\hat{T}} |\phi(t + p(x - \omega_1), x + qa(x - \omega_1))|^2 dtdx \qquad (7.9)$$

we perform the change of variables

$$\xi = t + p(x - \omega_1), \quad \eta = x + qa(x - \omega_1). \qquad (7.10)$$

In these variables, (7.9) is written as

$$(1 + qa) \int \int_{\Omega_{p,q}} |\phi(\xi, \eta)|^2 d\xi d\eta,$$

where $\Omega_{p,q}$ is the image of $(0, \hat{T}) \times (\omega_1, \hat{\omega}_2)$ by the mapping defined by (7.10). Besides, in view of (7.7), for all $(p, q) \in S$,

$$\Omega_{p,q} \subset (0, T) \times (\omega_1, \omega_2).$$

Thus,

$$\int_{\omega_1}^{\omega_2} \int_0^T |\phi(\xi, \eta)|^2 d\xi d\eta \geq \int \int_{\Omega_{p,q}} |\phi(\xi, \eta)|^2 d\xi d\eta.$$

This fact, in account of the inequality (7.8), proves the proposition.

The following proposition shows how the operators $\mathbf{B}_a$ act on the functions of the form $e^{\frac{in\pi t}{\rho\ell}} \sin(n\pi x/\ell)$. It is proved by simple calculations.

Proposition 7.5. *For all $\rho, \ell \in \mathbb{R}$ and $n \in \mathbb{N}$ the following identity holds*

$$\mathbf{B}_a \left(e^{\frac{in\pi t}{\rho\ell}} \sin(n\pi x/\ell) \right)$$

$$= -4e^{\frac{in\pi}{\rho\ell}(t + x - \omega_1)} \sin\frac{n\pi x}{\ell} \sin\left(\frac{n\pi(x - \omega_1)}{2\ell}\alpha\right) \sin\left(\frac{n\pi(x - \omega_1)}{2\ell}\beta\right),$$

where $\alpha = \left(\rho^{-1} + a\right)$ and $\beta = \left(\rho^{-1} - a\right)$.

Remark 7.6. If $\phi(t, x)$ is a solution of the wave equation

$$\rho^2 \phi_{tt} - \phi_{xx} = 0, \qquad \phi(t, 0) = \phi(t, \ell) = 0,$$

whose initial data $\phi|_{t=0}$ and $\phi_t|_{t=0}$ are finite linear combinations of the eigenfunctions $(\sin(n\pi x/\ell))$ then

$$\mathbf{B}_{\rho^{-1}} \phi(t, x) = 0.$$

Proposition 7.7. *Let $\ell, \alpha \neq \beta$ be positive numbers and I be an interval in $\mathbb{R}$. Then, there exists a constant $C > 0$ such that, for all $n \in \mathbb{R}$,*

$$\int_I \left| \sin\frac{n\pi x}{\ell} \sin\left(\frac{n\pi(x - \omega_1)}{2\ell}\alpha\right) \cdot \sin\left(\frac{n\pi(x - \omega_1)}{2\ell}\beta\right) \right|^2 dx \geq C.$$

This fact may be easily proved by computing the integral.

Proof (Proof of the Theorem 7.1). Let $(a_n^1)_{n\in\mathbb{Z}_*}$, $(a_n^2)_{n\in\mathbb{Z}_*}$ be complex finite sequences satisfying $a_{-n}^1 = \overline{a_n^1}$, $a_{-n}^2 = \overline{a_n^2}$ and

$$\phi^k(t,x) = \sum_{n\in\mathbb{Z}_*} a_n^k e^{\frac{in\pi t}{\rho_k \ell_k}} \sin(n\pi x/\ell_k), \quad k = 1, 2.$$

Let us take $a = \rho_1^{-1}$. Since $T > T_0$, it is possible to choose $\hat{\omega}_2 > \omega_1$ sufficiently close to ω_1 such that $\hat{\omega}_2$ and $\hat{T}$ satisfy (7.7) and, besides, $\hat{T} \geq T_0$. Then, according to Proposition 7.4,

$$C \int_{\omega_1}^{\omega_2} \int_0^T |\phi^1 + \phi^2|^2 dtdx \geq \int_{\omega_1}^{\hat{\omega}_2} \int_0^{\hat{T}} |\mathbf{B}_a\phi^1 + \mathbf{B}_a\phi^2|^2 dtdx. \tag{7.11}$$

But, from Remark 7.6,

$$\mathbf{B}_a\phi^1 = 0.$$

Thus, from inequality (7.11) it follows

$$C \int_{\omega_1}^{\omega_2} \int_0^T |\phi^1 + \phi^2|^2 dtdx \geq \int_{\omega_1}^{\hat{\omega}_2} \int_0^{\hat{T}} |\mathbf{B}_a\phi^2|^2 dtdx. \tag{7.12}$$

On the other hand, as

$$\phi^2(t,x) = \sum_{n\in\mathbb{Z}_*} a_n^2 e^{\frac{in\pi}{\rho_2 \ell_2}t} \sin(n\pi x/\ell_2),$$

Proposition 7.5 guarantees that

$$\mathbf{B}_a\phi^2(t,x) = \sum_{n\in\mathbb{Z}_*} a_n^2 e^{\frac{in\pi}{\rho_2 \ell_2}t} \Theta_n(x),$$

where

$$\Theta_n(x) := -4e^{\frac{in\pi}{\rho_2 \ell_2}(x-\omega_1)} \sin\frac{n\pi x}{\ell_2} \sin\left(\frac{n\pi(x-\omega_1)}{2\ell_2}\alpha\right) \sin\left(\frac{n\pi(x-\omega_1)}{2\ell_2}\beta\right)$$

with

$$\alpha = \frac{1}{\rho_2} + \frac{1}{\rho_1}, \qquad \beta = \frac{1}{\rho_2} - \frac{1}{\rho_1}.$$

Moreover, in view of Proposition 7.7, there exists a constant $C > 0$ such that for every $n \in \mathbb{N}$ the following inequality is verified.

$$\int_{\omega_1}^{\hat{\omega}_2} |\Theta_n(x)|^2 dx \geq C. \tag{7.13}$$

Therefore, since $\hat{T} \geq T_0 \geq 2\rho_2\ell_2$,

$$\int_{\omega_1}^{\hat{\omega}_2} \int_0^{\hat{T}} |\mathbf{B}_a \phi^2|^2 dt dx \geq \int_{\omega_1}^{\hat{\omega}_2} \int_0^{2\rho_2 \ell_2} \left| \sum_{n \in \mathbb{Z}_*} a_n^2 e^{\frac{in\pi}{\rho_2 \ell_2}t} \Theta_n(x) \right|^2 dt dx$$

$$= 2 \sum_{n \in \mathbb{N}} |a_n^2|^2 \int_{\omega_1}^{\hat{\omega}_2} |\Theta_n(x)|^2 dx,$$

(we have used the fact that the functions $e^{\frac{in\pi}{\rho_2 \ell_2}t}$ are orthogonal on $(0, 2\rho_2\ell_2)$) and then, in view of (7.12) and (7.13),

$$C \int_{\omega_1}^{\omega_2} \int_0^T |\phi^1 + \phi^2|^2 dt dx \geq \sum_{n \in \mathbb{N}} |a_n^2|^2.$$

Inequality

$$C \int_{\omega_1}^{\omega_2} \int_0^T |\phi^1 + \phi^2|^2 dt dx \geq \sum_{n \in \mathbb{N}} |a_n^1|^2$$

is proved in a similar way, applying to $\phi^1 + \phi^2$ the operator $\mathbf{B}_a$ with $a = \rho_2^{-1}$. This concludes the proof of the theorem.

7.1.3 Control of Strings with Equal Densities

Theorem 7.1 does not provide any information on what happens when $\rho_1 = \rho_2$ but $\ell_1 \neq \ell_2$. This is due to the local character of the operators $\mathbf{B}_a$: they cannot distinguish between solutions of the wave equation that propagate at the same speed. This fact, however, is not purely technical. If $\rho_1 = \rho_2 = \rho$, the condition $\ell_1 \neq \ell_2$ is not sufficient for the inequality (7.6) to be true.

Indeed, let us assume that

$$\frac{\ell_1}{\ell_2} = \frac{p}{q}, \quad p, q \in \mathbb{N}.$$

Then, the solutions

$$\phi^1(t, x) = e^{\frac{ip\pi}{\rho\ell_1}t} \sin \frac{p\pi x}{\ell_1}, \quad \phi^2(t, x) = -e^{\frac{iq\pi}{\rho\ell_2}t} \sin \frac{q\pi x}{\ell_2}$$

satisfy

$$\phi^1(t, x) + \phi^2(t, x) \equiv 0.$$

Thus, an inequality of type (7.6) is impossible for any interval ω and any time T, whenever the ratio ℓ_1/ℓ_2 is a rational number. It is even impossible to replace the right hand term in (7.6) by any other weaker norm of the initial data. In this sense, the problem turns out to be similar to that of the simultaneous control of two strings from one extreme, since the lengths of the strings do play a crucial role. When the ratio ℓ_1/ℓ_2 is an irrational number, it is possible to prove a weakened version of (7.6). We use the same technique as in Theorem 4.3.

We denote by Z^k, $k = 1, 2$, the space of the finite linear combinations of the functions $\left(\sin \frac{n\pi x}{\ell_k} \right)_{n \in \mathbb{N}}$.

Theorem 7.8. *Let $\rho_1 = \rho_2 = \rho$ and $T \geq 2\rho(\ell_1 + \ell_2)$. Then there exists a constant $C > 0$ such that*

$$C \int_\omega \int_0^T |\phi^1(t,x) + \phi^2(t,x)|^2 dt dx \geq \sum \sin^2 \frac{\ell_2 n\pi}{\ell_1} \left((\phi_0^1)^2 + n^{-2}(\phi_1^1)^2 \right),$$

$$(7.14)$$

$$C \int_\omega \int_0^T |\phi^1(t,x) + \phi^2(t,x)|^2 dt dx \geq \sum \sin^2 \frac{\ell_1 n\pi}{\ell_2} \left((\phi_0^2)^2 + n^{-2}(\phi_1^2)^2 \right),$$

$$(7.15)$$

for all the solutions ϕ^1, ϕ^2 of (7.4) with initial states in $Z^1 \times Z^1$ and $Z^2 \times Z^2$, respectively.

Proof. The inequality (7.15) is equivalent to

$$C \int_\omega \int_0^T |\Phi|^2 \, dt dx \geq \sum_{n \in \mathbb{N}} |a_n^2|^2 \sin^2 \frac{\ell_1 n\pi}{\ell_2},$$

where

$$\Phi := \sum_{n \in \mathbb{Z}_*} a_n^1 e^{\frac{in\pi t}{\rho\ell_1}} \sin \frac{n\pi x}{\ell_1} + a_n^2 e^{\frac{in\pi t}{\rho\ell_2}} \sin \frac{n\pi x}{\ell_2},$$

for all the finite complex sequences verifying $a_{-n}^1 = \overline{a_n^1}$, $a_{-n}^2 = \overline{a_n^2}$.

To prove this assertion, let us observe that, for every $x \in \omega$,

$$(\rho\ell_1)^- \left(\sum_{n \in \mathbb{Z}_*} a_n^1 e^{\frac{in\pi t}{\rho\ell_1}} \right) = 0,$$

$$(7.16)$$

where $(\rho\ell_1)^-$ is the operator defined by (3.7) for the number $\rho\ell_1$ (the equality (7.16) corresponds to the $2\rho\ell_1$-periodicity in time of the solution ϕ^1). Then,

$$(\rho\ell_1)^- \Phi = (\rho\ell_1)^- \sum_{n \in \mathbb{Z}_*} a_n^2 e^{\frac{in\pi t}{\rho\ell_2}} \sin \frac{n\pi x}{\ell_2}$$

$$= \sum_{n \in \mathbb{Z}_*} a_n^2 e^{\frac{in\pi t}{\rho\ell_2}} \sin \frac{\ell_1 n\pi}{\ell_2} \sin \frac{n\pi x}{\ell_2}.$$

Besides, from Proposition 3.3 we obtain that for every $x \in \omega$,

$$\int_0^T |\Phi|^2 \, dt \geq \int_{\rho\ell_1}^{T-\rho\ell_1} |(\rho\ell_1)^- \Phi|^2 dt$$

$$= \int_{\rho\ell_1}^{T-\rho\ell_1} \left| \sum_{n \in \mathbb{Z}_*} a_n^2 e^{\frac{in\pi t}{\rho\ell_2}} \sin \frac{\ell_1 n\pi}{\ell_2} \sin \frac{n\pi x}{\ell_2} \right|^2 dt. \quad (7.17)$$

On the other hand, since $T \geq 2\rho(\ell_1 + \ell_2)$ then,

$$\int_{\rho\ell_1}^{T-\rho\ell_1} \left| \sum_{n\in\mathbb{Z}_*} a_n^2 e^{\frac{in\pi t}{\rho\ell_2}} \sin\frac{\ell_1}{\ell_2}n\pi \sin\frac{n\pi x}{\ell_2} \right|^2 dt$$

$$\geq \int_{\rho\ell_1}^{\rho\ell_1+2\rho\ell_2} \left| \sum_{n\in\mathbb{Z}_*} a_n^2 e^{\frac{in\pi t}{\rho\ell_2}} \sin\frac{\ell_1 n\pi}{\ell_2} \sin\frac{n\pi x}{\ell_2} \right|^2 dt$$

$$= 2\sum_{n\in\mathbb{N}} |a_n^2|^2 \sin^2\frac{n\pi x}{\ell_2} \sin^2\frac{\ell_1 n\pi}{\ell_2}.$$

Here we have used the fact that the functions $\left(e^{\frac{in\pi t}{\rho\ell_2}} \right)_{n\in\mathbb{Z}_*}$ are orthonormal on any interval of length $2\rho\ell_2$.

Further, in view of (7.17),

$$C \int_\omega \int_0^T |\Phi|^2 \, dt dx \geq \sum_{n\in\mathbb{N}} |a_n^2|^2 \sin^2\frac{\ell_1 n\pi}{\ell_2} \int_\omega \sin^2\frac{n\pi x}{\ell_2} dx. \qquad (7.18)$$

Finally, let us observe that for any interval $\omega \subset \mathbb{R}$ there exists a constant $C = C(\omega)$ such that

$$\int_\omega \sin^2\frac{n\pi x}{\ell_2} dx \geq C.$$

Therefore, from (7.18) it holds

$$C \int_\omega \int_0^T |\Phi|^2 \, dt dx \geq \sum_{n\in\mathbb{N}} |a_n^2|^2 \sin^2\frac{\ell_1 n\pi}{\ell_2}.$$

The inequality (7.14) may be obtained in an analogous way.

Remark 7.9. When the number ℓ_1/ℓ_2 is rational, some of the coefficients $\sin(\ell_2 n\pi/\ell_1)$ or $\sin(\ell_1 n\pi/\ell_2)$ entering in the right hand side term of inequalities (7.14), (7.15) vanish. This agrees with the fact that, in this case, we cannot obtain an inequality of type (7.6).

Corollary 7.10. *If the number ℓ_1/ℓ_2 is irrational and $T \geq 2\rho(\ell_1 + \ell_2)$ then system (7.2) is spectrally controllable in time T, that is, all the initial states $(u_0^1, u_1^1) \in Z^1 \times Z^1$, $(u_0^2, u_1^2) \in Z^2 \times Z^2$ are controllable in time T.*

If we have some additional information on the rational approximation properties of the ratio ℓ_1/ℓ_2, then it is possible to describe subspaces of controllable initial states in the same way as it was done in Subsection 4.2.1.

Corollary 7.11. *a) If $\ell_1/\ell_2 \in \mathbf{B}_\varepsilon$ then the subspace of initial states*

$$\hat{H}^{2+\varepsilon}(0,\ell_i) \times \hat{H}^{1+\varepsilon}(0,\ell_i),$$

is controllable in any time $T \geq 2\rho(\ell_1 + \ell_2)$. In particular, if ℓ_1/ℓ_2 is an irrational algebraic number, this subspace is controllable for any $\varepsilon > 0$.

b) If ℓ_1/ℓ_2 admits a bounded development in continuous fractions, then the subspace of initial states $\left[H^2(0,\ell_i) \cap H_0^1(0,\ell_i) \right] \times H_0^1(0,\ell_i)$, is controllable in any time $T \geq 2\rho(\ell_1 + \ell_2)$.

7.2 Simultaneous Control on the Whole Domain

Let Ω be a bounded open subset of $\mathbb{R}^n$ with smooth boundary and $f \in L^2_{\text{loc}}(\mathbb{R}^{n+1})$. Let us consider the system

$$\begin{cases} \rho_k^2 u_{tt}^k - \Delta u^k + f = 0 & \text{in } \Omega \times \mathbb{R}, \\ u^k |_{\partial \Omega} = 0 & \text{in } \mathbb{R}, \\ u^k(0,.) = u_0^k, \quad u_t^k(0,.) = u_1^k & \text{in } \Omega. \end{cases} \quad (7.19)$$

It is a model for the motion of N elastic membranes with densities $\rho_1, ..., \rho_N$ having at rest the same shape Ω and whose borders are fixed. Those membranes are controlled by means of a function f that acts on the whole domain Ω. When $n = 1$ the system (7.19) is a particular case of (7.2) with $\ell_1 = \ell_2 = \ell$ and $\omega = (0, \ell)$.

The problem (7.19) is well posed for initial states $(u_0^k, u_1^k) \in H_0^1(\Omega) \times L^2(\Omega)$, $k = 1, ..., N$. When $f = 0$, (7.19) becomes the homogeneous system

$$\begin{cases} \rho_k^2 \phi_{tt}^k - \Delta \phi^k = 0 & \text{in } \Omega \times \mathbb{R}, \\ \phi^k |_{\partial \Omega} = 0 & \text{in } \mathbb{R}, \\ \phi^k(0,.) = \phi_0^k, \quad \phi_t^k(0,.) = \phi_1^k & \text{in } \Omega, \end{cases} \quad (7.20)$$

which is also well posed for initial states $(\phi_0^k, \phi_1^k) \in L^2(\Omega) \times H^{-1}(\Omega)$, $k = 1, ..., N$.

If $(\mu_n)_{n \in \mathbb{N}}$ is the increasing sequence of the eigenvalues $-\Delta$ with Dirichlet homogeneous boundary conditions in Ω and $(\theta_n)_{n \in \mathbb{N}}$ the corresponding eigenfunctions, which constitute an orthonormal basis of $L^2(\Omega)$, then the solutions of (7.20) are determined by the formulas

$$\phi^k(t, x) = \sum_{n \in \mathbb{Z}_*} a_n^k e^{\frac{\lambda_n t}{\rho_k}} \theta_{|n|}(x)$$

where

$$\lambda_n = \sqrt{\mu_{|n|}} \, \text{sgn} \, n, \quad n \in \mathbb{Z}_*,$$

$$a_n^k = \frac{1}{2} \left(\phi_{0,|n|}^k - i \lambda_n \phi_{1,|n|}^k \right), \quad k = 1, ..., N, \quad n \in \mathbb{Z}_*.$$

The control problem associated to system (7.19) is: *given $T > 0$, to determine for which initial states $(u_0^k, u_1^k) \in H_0^1(\Omega) \times L^2(\Omega)$, $k = 1, ..., N$, there exists $f \in L^2((0,T) \times \Omega)$ such that the solution of (7.19) satisfies*

$$u^k |_{t=T} = u_t^k |_{t=T} = 0.$$

System (7.19) is said to be *exactly controllable in time T* when all the initial states from $H_0^1(\Omega) \times L^2(\Omega)$ are controllable in time T.

The application of HUM guarantees that system (7.19) is exactly controllable in time T if, and only if, there exists a constant $C > 0$ such that the inequality

$$C \int_{\Omega} \int_0^T \left| \sum_{k=1}^N \phi^k(t,x) \right|^2 dt dx \geq \sum_{k=1}^N \|(\phi_0^k, \phi_1^k)\|^2_{L^2(\Omega) \times H^{-1}(\Omega)} \qquad (7.21)$$

is verified by all the solutions of (7.20) with initial states $(\phi_0^k, \phi_1^k) \in L^2(\Omega) \times H^{-1}(\Omega)$, $k = 1, ..., N$.

This fact is equivalent to the existence of a constant $C > 0$ such that

$$C \int_{\Omega} \int_0^T \left| \sum_{k=1}^N \sum_{n \in \mathbb{Z}_*} a_n^k e^{\frac{\lambda_n t}{\rho_k}} \theta_{|n|}(x) \right|^2 dt dx \geq 2 \sum_{k=1}^N \sum_{n \in \mathbb{N}} |a_n^k|^2, \qquad (7.22)$$

for all the finite complex sequences $(a_n^k)_{n \in \mathbb{Z}_*}$, $k = 1, ..., N$, verifying $a_{-n}^k = \overline{a_n^k}$.

Theorem 7.12. *System (7.19) is exactly controllable in time $T > 0$ if, and only if, the numbers $\rho_1, ..., \rho_N$ are pairwise distinct.*

Remark 7.13. This is a simple result on the simultaneous controllability of a finite number of wave equations with different velocities of propagation in the same domain. The issue of simultaneous controllability was analyzed for the first time by J. L. Lions in [78], but this problem was not discussed in that monograph.

The elementary arguments we use in the proof of this result do not allow addressing the same problem when the subdomain of control is restricted to be a subset ω of the whole domain Ω. In order to get a sharp necessary and sufficient condition for the control in that case, microlocal tools need to be used. It is clear that, a necessary condition for controllability is that the subset ω where the control is being applied, satisfies the so called Geometric Control Condition (GCC) (see [18]) which asserts that all rays of Geometric Optics propagating in Ω and being reflected on its boundary enter the control region ω in a uniform time. One may expect that, under this GCC, when all densities are different, the exact controllability of the system will hold in a time which is the time given by this microlocal argument: the time that rays need to get into ω according to the slowest velocity of propagation of the different wave equations involved in the system. This would be a natural extension of the results above in one space dimension. But a complete proof of this result needs to be developed. Note that when the densities of two of the wave equations entering in the are the same, the exact controllability and approximate controllability properties fail, even in the simplest case where the control is applied everywhere in the domain. The fact that the densities are different plays a key role that, at the microlocal level, is reflected on the fact that the characteristic manifolds have empty intersection.

At this respect the work by N. Burq and G. Lebeau [25] is worth mentioning. There, the microlocal tools in [18] are further developed to analyze the propagation of the polarization of singularities for systems of wave equations. These tools will definitely be extremely useful to address problems of simultaneous controllability and, more generally, to analyse observability and controllability properties of systems of wave equations on networks.

There is an extensive literature on the controllability and stabilization of weakly coupled systems of wave equations (see, for instance, [1]). There, the wave equations entering in the system are considered to have the same principal part and to be coupled through lower order terms. Obviously this coupling may produce the approximate controllability or even controllability with a loss of a finite number of derivatives in the space of controllable data. But this is issue, as far as we know, is widely open.

Proof. If two of the numbers $\rho_1, ..., \rho_N$ coincide, say $\rho_1 = \rho_2$, and we choose

$$a_n^1 = -a_n^2, \quad a_n^k = 0, \quad k > N,$$

then the inequality (7.22) becomes

$$0 \geq 4 \sum_{n \in \mathbb{N}} |a_n^1|^2,$$

what is not true in general. Therefore, if two of the numbers $\rho_1, ..., \rho_N$ coincide, (7.22) fails.

Let us observe that, due to the orthonormality in $L^2(\Omega)$ of the functions $(\theta_n)_{n \in \mathbb{N}}$, the inequality (7.22) may be written as

$$C \int_0^T \sum_{n \in \mathbb{N}} \left| \sum_{k=1}^N a_n^k e^{\frac{\lambda_n t}{\rho_k}} \right|^2 dt \geq \sum_{k=1}^N \sum_{n \in \mathbb{N}} |a_n^k|^2.$$

Then, for every $T > 0$ and distinct numbers $\rho_1, ..., \rho_N$ it suffices to apply Proposition 7.14 given below to obtain (7.22) and consequently, the proof of the theorem.

Proposition 7.14. *Let $\rho_1, ..., \rho_N$ be distinct positive numbers and $(\lambda_n)_{n \in \mathbb{N}}$ a sequence of positive numbers that tends to infinite. Then, for every $T > 0$ there exists a constant $C = C(T, N, \rho_1, ..., \rho_N) > 0$ such that*

$$\int_0^T \left| \sum_{k=1}^N a^k e^{\frac{i\lambda_n t}{\rho_k}} \right|^2 dt \geq C \sum_{k=1}^N |a^k|^2,$$

for all $n \in \mathbb{N}$ and $(a^1, ..., a^N) \in \mathbb{R}^N$.

Proof. We proceed by induction with respect to the number N. For $N = 1$ the inequality is immediate. Let us suppose that the inequality is true for $N - 1$.

Let us denote

$$I_n = \sum_{k=2}^{N} a^k e^{\frac{i\lambda_n t}{\rho_k}}.$$

Then, according to our induction hypothesis, there exists a constant $C > 0$ such that, for every $n \in \mathbb{N}$,

$$\int_0^T |I_n|^2\, dt \geq C \sum_{k=2}^{N} |a^k|^2.$$

On the other hand,

$$\int_0^T \left| \sum_{k=1}^{N} a^k e^{\frac{i\lambda_n t}{\rho_k}} \right|^2 dt = |a^1|^2 T + 2\Re\left(\int_0^T a^1 e^{\frac{i\lambda_n t}{\rho_1}} \overline{I_n} dt \right) + \int_0^T |I_n|^2 dt. \quad (7.23)$$

Let us observe that

$$\left| \Re \int_0^T a^1 e^{\frac{i\lambda_n t}{\rho_1}} \overline{I_n} dt \right| \leq |a^1| \left| \int_0^T e^{\frac{i\lambda_n t}{\rho_1}} \overline{I_n} dt \right| = |a^1| \left| \sum_{k=2}^{N} a^k \gamma_{n,k} \right|, \quad (7.24)$$

where

$$\gamma_{n,k} = \int_0^T e^{\left(\frac{1}{\rho_1} - \frac{1}{\rho_k} \right) i\lambda_n t} dt.$$

Besides,

$$|a^1| \left| \sum_{k=2}^{N} a^k \gamma_{n,k} \right| \leq |a^1| \sum_{k=2}^{N} |a^k| |\gamma_{n,k}| \leq \frac{1}{2} \sum_{k=2}^{N} \left(|a^1|^2 + |a^k|^2 \right) |\gamma_{n,k}|. \quad (7.25)$$

Combining (7.24) and (7.25) it holds

$$\left| \Re \int_0^T a^1 e^{\frac{i\lambda_n t}{\rho_1}} \overline{I_n} dt \right| \leq \frac{1}{2} \sum_{k=2}^{N} \left(|a^1|^2 + |a^k|^2 \right) |\gamma_{n,k}|,$$

which, in view of (7.23), implies

$$\int_0^T \left| \sum_{k=1}^{N} a^k e^{\frac{i\lambda_n t}{\rho_k}} \right|^2 dt \geq |a^1|^2 \left(T - \sum_{k=2}^{N} |\gamma_{n,k}| \right) + \sum_{k=2}^{N} |a^k|^2 (C - |\gamma_{n,k}|). \quad (7.26)$$

Let us observe now that[2], for every $k = 2, ..., N$,

$$|\gamma_{n,k}| \leq \frac{2|\rho_k - \rho_1|}{\rho_1 \rho_k \lambda_n} \underset{n \to \infty}{\longrightarrow} 0.$$

---

[2] It is precisely at this point of the proof where the condition that the numbers ρ_k, $k = 1, ..., N$, are all distinct is esential.

Therefore, there exists $n_0 \in \mathbb{N}$ such that, for all $n \geq n_0$,

$$\sum_{k=2}^{N} |\gamma_{n,k}| \leq \frac{T}{2}, \qquad |\gamma_{n,k}| \leq \frac{C}{2}, \quad k = 2, ..., N.$$

As a consequence of (7.26) it holds, for every $n \geq n_0$,

$$\int_0^T \left| \sum_{k=1}^{N} a^k e^{\frac{i\lambda_n t}{\rho_k}} \right|^2 dt \geq |a^1|^2 \frac{T}{2} + \frac{C}{2} \sum_{k=2}^{N} |a^k|^2 \geq C \sum_{k=1}^{N} |a^k|^2. \qquad (7.27)$$

Finally, it suffices to note that the functions $e^{i\lambda_n t/\rho_k}$, $k = 1, ..., N$, are linearly independent over any interval, and thus

$$\int_0^T \left| \sum_{k=1}^{N} a^k e^{\frac{i\lambda_n t}{\rho_k}} \right|^2 dt > 0,$$

except when $a^1 = \cdots = a^N = 0$. This allows to apply a standard compactness argument to prove that, for every $n \in \mathbb{N}$, there exists a constant $C_n > 0$ such that

$$\int_0^T \left| \sum_{k=1}^{N} a^k e^{\frac{i\lambda_n t}{\rho_k}} \right|^2 dt \geq C_n \sum_{k=1}^{N} |a^k|^2.$$

Therefore, there exists $C > 0$ such that

$$\int_0^T \left| \sum_{k=1}^{N} a^k e^{\frac{i\lambda_n t}{\rho_k}} \right|^2 dt \geq C \sum_{k=1}^{N} |a^k|^2,$$

for every $n < n_0$. This fact, in view of (7.27), gives the assertion of the proposition.

8

Other Equations on Networks

In this chapter we study observation and control problems for the heat, beam and Schrödinger equations on networks. We make emphasis on two main issues: the spectral observability/controllability of the corresponding systems and the possibility of identifying subspaces of controllable initial data. Our main tool consists in using the information we have obtained in previous chapters on the controllability of the wave-system (2.11)-(2.16) on the same network, and spectral transformations allowing to transfer that information into those other systems.

The main spectral controllability result that we present in this chapter asserts that, whenever the system (2.11)-(2.16) is spectrally controllable in some time $T > 0$, the heat, beam and Schrödinger equations are also spectrally controllable in any time $\tau > 0$. This result is in agreement with the already existing ones on the control of those systems in domains of $\mathbb{R}^n$ in any dimension. Indeed, it is well known that those systems have better controllability properties than the wave equation in the sense that, whenever the wave equation is controllable, the other models are controllable too but in an arbitrarily small time. The later fact is due to the intrinsic infinite velocity of propagation of these other models.

Then in our setting, according to these results and in view of Theorem 6.5, the spectral controllability of those systems (heat, beam and Schrödinger equations) admits the following characterization: these systems are spectrally controllable in any time $\tau > 0$ if, and only if, no eigenfunction of the elliptic operator $-\Delta_G$ associated to the system (2.11)-(2.16) vanishes identically on the controlled string.

The results on system (2.11)-(2.16) (corollaries 5.38 and 5.39) for networks of strings allowing to characterize, under suitable assumptions, subspaces of controllable data, will allow us to identify subspaces of controllable data of the form V^r (domains of powers of the operator $-\Delta_G$) for these other systems too.

8.1 The Heat Equation

The following parabolic system will be called heat equation on a network:

$$u_t^i - u_{xx}^i = 0 \qquad \text{in } \mathbb{R} \times [0, \ell_i], \quad i = 1, ..., M, \tag{8.1}$$

$$u^1(t, \mathbf{v}_1) = h(t) \qquad t \in \mathbb{R}, \tag{8.2}$$

$$u^{i(j)}(t, \mathbf{v}_j) = 0 \qquad t \in \mathbb{R}, \quad j = 2, ..., N, \tag{8.3}$$

$$u^i(t, \mathbf{v}) = u^j(t, \mathbf{v}) \qquad t \in \mathbb{R}, \quad \mathbf{v} \in \mathcal{V}_M, \, i, j \in I_{\mathbf{v}}, \tag{8.4}$$

$$\sum_{i \in I_{\mathbf{v}}} \partial_n u^i(t, \mathbf{v}) = 0 \qquad t \in \mathbb{R}, \, \mathbf{v} \in \mathcal{V}_M, \tag{8.5}$$

$$u^i(0, x) = u_0^i(x) \qquad x \in [0, \ell_i], \quad i = 1, ..., M. \tag{8.6}$$

The problem (8.1)-(8.6) may be viewed as a model for heat propagation in a network under the action of a controller on one of the exterior nodes. For every $T > 0$, $h \in L^2(0,T) \cap C([0,T])$ [1] and $\bar{u}_0 = (u_0^1, ..., u_0^M) \in H$ system (8.1)-(8.6) has a unique solution $\bar{u}$ that satisfies

$$\bar{u} \in C([0,T] : H).$$

When $h \equiv 0$, this solution is expressed by the formula

$$\bar{u}(t, x) = \sum_{n \in \mathbb{N}} u_{0,n} e^{-\mu_n t} \bar{\theta}_n(x),$$

if $\bar{u}_0 = \sum_{n \in \mathbb{N}} u_{0,n} \bar{\theta}_n$. Recall that $\mu_n = \lambda_n^2$ are the eigenvalues and $\bar{\theta}_n$ the eigenfunctions of the Dirichlet problem for the laplacian on the network, which is the same that corresponds to the wave-like model (2.11)-(2.15).

For system (8.1)-(8.6) we consider the following control problem: *To determine for which initial data $\bar{u}_0 \in V'$, there exists a function $h \in L^2(0,T)$ such that the solution $\bar{u}$ of (8.1)-(8.6) satisfies*

$$\bar{u}(T, x) = \bar{0}.$$

When the initial datum $\bar{u}_0$ has this property it is said that $\bar{u}_0$ is *controllable to zero in time T*. If all the initial data $\bar{u}_0 \in Z$ are controllable to zero (or null-controllable) in time T (as before, Z is the set of all the finite linear combinations of the eigenfunctions), we will say that the system (8.1)-(8.6) is spectrally controllable in time T.

---

[1] When one simply assumes that $h \in L^2(0,T)$ the regularity of the solutions of the heat equation is slightly weaker, say, $u \in C([0,T] : H^{-1})$. But this suffices to define the problem of null control, since the value of the solution at time $t = T$ is well defined. This is in practice irrelevant since the controls one obtains are related to traces of solutions of the adjoint heat equation and therefore are smooth. We refer to [81] for a careful analysis of this issue in the context of the homogenization of the heat equation.

Let us observe that, unlike it happens for the wave equation, or more generally, for time-reversible equations, the fact that $\bar{u}_0$ and $\bar{u}_1$ are controllable to zero does not imply the existence of a function $h \in L^2(0,T)$ such that the solution of (8.1)-(8.6) with initial datum $\bar{u}_0$ coincides with $\bar{u}_1$ in time T. This is due to the strong time irreversibility of the heat equation: due to the dissipative character of the heat operator, the solutions with initial data $\bar{u}_0 \in H$ satisfy $u^i(T) \in C^\infty((0, \ell_i))$ for $i = 2, ..., M$. Thus, only very smooth states of the system may be reached. [2]

Proceeding as in the case of the wave equation, we obtain the following criterion for null-controllability:

Proposition 8.1. *The initial datum $\bar{u}_0 \in H$ is controllable to zero in time T with control $h \in L^2(0,T)$ if, and only if, for every $\bar{\phi}_0 \in Z$ the following identity holds*

$$\langle \bar{u}_0, \bar{\phi}(T) \rangle_H = \int_0^T h(t) \partial_n \phi^1(T - t, \mathbf{v}_1) dt,$$

where $\bar{\phi}$ is the solution of the homogeneous system (8.1)-(8.6) with initial datum $\bar{\phi}_0$.

Clearly, it is in fact sufficient to check the equality of the previous proposition when $\bar{\phi}_0$ is one of the eigenfunctions. That is,

$$\int_0^T \varkappa_k e^{-\mu_k(T-t)} h(t) dt = u_{0,k} e^{-\mu_k T}, \qquad k \in \mathbb{N}, \tag{8.7}$$

where $\varkappa_k = \partial_n \theta_k^1(\mathbf{v}_1)$ is the normal derivative of the eigenfunction $\bar{\theta}_k$ at the controlled node $\mathbf{v}_1$.

Accordingly, performing the change of variable $t \to T/2 - t$, the control problem may be written equivalently as the following problem of moments:

Proposition 8.2. *The initial datum $\bar{u}_0 = \sum_{n \in \mathbb{N}} u_{0,n} \bar{\theta}_n \in H$ is controllable to zero in time T with control $h \in L^2(0,T)$ if, and only if, the following identities are satisfied*

$$\int_{-T/2}^{T/2} \varkappa_k e^{-\mu_k t} h(t) dt = u_{0,k} e^{-\mu_n T/2}, \qquad k \in \mathbb{N}. \tag{8.8}$$

This proposition allows giving the following characterization of the networks for which system (8.1)-(8.6) is spectrally controllable.

Theorem 8.3. *System (8.1)-(8.6) is spectrally controllable to zero in any time $T > 0$ if, and only if, $\varkappa_k \neq 0$ for every $k \in \mathbb{N}$.*

---

[2] Essentially, those in the range of the uncontrolled semigroup, starting from the subspace of controllable initial data.

Proof. The necessity of the condition $\varkappa_k \neq 0$ is immediate: if $\varkappa_k = 0$ for some value of k then, the equality (8.8) becomes $u_{0,k} = 0$. Consequently, it will not be possible to control an initial datum $\bar{u}_0 = \bar{\theta}_k \in Z$ with $u_{0,k} = 1$.

In order to prove that the condition $\varkappa_k \neq 0$ for every $k \in \mathbb{N}$ is sufficient for the spectral controllability to zero of system (8.1)-(8.6), it is enough to prove that for every $T > 0$ there exists a sequence (w_n), which is biorthogonal to $(\varkappa_k e^{-\mu_k t})$ in $L^2(-T/2, T/2)$.

According to Theorem 6.5 system (2.11)-(2.16) is spectrally controllable in time $T = 2L$ (recall that L is the total length of the graph). Then, using Proposition 3.24, there exists a sequence $(v_n)_{n \in \mathbb{Z}_*}$ biorthogonal to $(\varkappa_k e^{i\lambda_k t})$ in $L^2(-L, L)$.

From Theorem 3.22 we conclude that for every $T > 0$ there exists a sequence (w_n) biorthogonal to $(\varkappa_k e^{-\mu_k t})$ in $L^2(-T/2, T/2)$.

Remark 8.4. When the graph of the network is a tree, the condition $\varkappa_k \neq 0$ for all $k \in \mathbb{N}$ coincides with the fact that any two sub-trees with common root have disjoint spectra. Recall that this is the non-degeneracy condition introduced in Chapter 5.

Following the procedure introduced by Russell in [105], it is possible to obtain additional information on the controllability of system (8.1)-(8.6) as a consequence of the controllability of subspaces of initial states of the form $\mathcal{W}^r$ for the network of strings:

Proposition 8.5. *If the subspace $\mathcal{W}^r$ is controllable for system (2.11)-(2.16) in time $T > 0$ for some finite r, then all the initial data $\bar{u}_0 \in H$ are controllable to zero in any time $\tau > 0$ for system (8.1)-(8.6).*

Proof. According to Proposition 3.25, if $\mathcal{W}^r$ is controllable for system (2.11)-(2.16) in time $T > 0$ then there exists a sequence $(v_n)_{n \in \mathbb{Z}_*}$ biorthogonal to $(\varkappa_k e^{i\lambda_k t})$ in $L^2(-T/2, T/2)$. Besides, there exists a constant $C > 0$ such that for every $n \in \mathbb{Z}_*$, the sequence (v_n) satisfies

$$\|v_n\|_{L^2(-T/2, T/2)} \leq C\lambda_n^{r-1}. \tag{8.9}$$

From Russell's Theorem 3.22 we obtain that for every $\tau > 0$ there exists a sequence (w_n) biorthogonal to $(\varkappa_k e^{-\mu_k t})$ in $L^2(-\tau/2, \tau/2)$, for which there exist positive constants C_τ and γ such that

$$\|w_n\|_{L^2(-\tau/2, \tau/2)} \leq C_\tau \|v_n\|_{L^2(-T/2, T/2)} e^{\gamma\lambda_n}, \tag{8.10}$$

for every $n \in \mathbb{N}$.

In view of (8.9), (8.10) we obtain

$$\|w_n\|_{L^2(-\tau/2, \tau/2)} \leq C\lambda_n^{r-1} e^{\gamma\lambda_n}. \tag{8.11}$$

Finally, applying Proposition 3.18 to the problem of moments (8.8) it follows that all the initial data $\bar{u}_0 \in H$ satisfying

$$\sum_{n \in \mathbb{N}} \left| u_{0,n} e^{-\mu_n T/2} \right| \|w_n\|_{L^2(-\tau/2, \tau/2)} < \infty \tag{8.12}$$

are controllable.

In view of (8.11), applying the Cauchy-Schwarz inequality we obtain that the convergence (8.12) is guaranteed if

$$\left(\sum_{n \in \mathbb{N}} |u_{0,n}|^2 \right) \left(\sum_{n \in \mathbb{N}} \lambda_n^{2r-2} e^{2\gamma \lambda_n - \mu_n T} \right) < \infty.$$

Since $\mu_n = \lambda_n^2$ and $\lambda_n \to \infty$, the series

$$\sum_{n \in \mathbb{N}} \lambda_n^{2r-2} e^{2\gamma \lambda_n - \mu_n T}$$

converges for any $r \in \mathbb{R}$ and then, all the initial data satisfying

$$\sum_{n \in \mathbb{N}} |u_{0,n}|^2 < \infty;$$

are controllable. Consequently, any $\bar{u}_0 \in H$ is controllable.

Theorem 8.3 and Proposition 8.5 allow to immediately obtain from corollaries 5.38 and 5.39, information on the controllability of the heat equation on the star-shaped networks studied in Section 5.8 of Chapter 5.

Corollary 8.6. *If the lengths $\ell_1, ..., \ell_{n-1}$ of the uncontrolled edges of the star-shaped network are such that*

1) *all the ratios ℓ_i/ℓ_j with $i \neq j$ are irrational numbers, then system (8.1)-(8.6) is spectrally controllable to zero in any time $T > 0$.*
2) *all the ratios ℓ_i/ℓ_j with $i \neq j$ belong to some set $\mathbf{B}_\varepsilon$, then all the initial data $\bar{u}_0 \in H$ are controllable to zero in any time $T > 0$.*

8.2 The Schrödinger Equation

Let us consider the Schrödinger system on the network:

$$iu_t^k - u_{xx}^k = 0 \qquad \text{in } \mathbb{R} \times [0, \ell_k], \quad k = 1, ..., M, \tag{8.13}$$

$$u^1(t, \mathbf{v}_1) = h(t) \qquad t \in \mathbb{R}, \tag{8.14}$$

$$u^{k(j)}(t, \mathbf{v}_j) = 0 \qquad t \in \mathbb{R}, \quad j = 2, ..., N, \tag{8.15}$$

$$u^k(t, \mathbf{v}) = u^j(t, \mathbf{v}) \qquad t \in \mathbb{R}, \quad \mathbf{v} \in \mathcal{V}_\mathcal{M}, \ k, j \in k_\mathbf{v}, \tag{8.16}$$

$$\sum_{k \in k_\mathbf{v}} \partial_n u^k(t, \mathbf{v}) = 0 \qquad t \in \mathbb{R}, \quad \mathbf{v} \in \mathcal{V}_\mathcal{M}, \tag{8.17}$$

$$u^k(0, x) = u_0^k(x) \qquad x \in [0, \ell_k], \quad k = 1, ..., M. \tag{8.18}$$

We consider also its homogeneous version:

$$i\phi_t^k - \phi_{xx}^k = 0 \qquad\qquad \text{in } \mathbb{R} \times [0, \ell_k], \quad k = 1, ..., M, \tag{8.19}$$

$$\phi^1(t, \mathbf{v}_1) = 0 \qquad\qquad t \in \mathbb{R}, \tag{8.20}$$

$$\phi^{k(j)}(t, \mathbf{v}_j) = 0 \qquad\qquad t \in \mathbb{R}, \quad j = 2, ..., N, \tag{8.21}$$

$$\phi^k(t, \mathbf{v}) = \phi^j(t, \mathbf{v}) \qquad\quad t \in \mathbb{R}, \quad \mathbf{v} \in \mathcal{V}_\mathcal{M}, \; k, j \in k_\mathbf{v}, \tag{8.22}$$

$$\sum_{k \in k_\mathbf{v}} \partial_n \phi^k(t, \mathbf{v}) = 0 \qquad t \in \mathbb{R}, \quad \mathbf{v} \in \mathcal{V}_\mathcal{M}, \tag{8.23}$$

$$\phi^k(0, x) = \phi_0^k(x) \qquad\qquad x \in [0, \ell_k], \quad k = 1, ..., M. \tag{8.24}$$

For every $T > 0$ and $\bar{\phi}_0 = (\phi_0^1, ..., \phi_0^M) \in V$ system (8.19)-(8.24) has a unique solution $\bar{\phi}$, which is expressed by the formula

$$\bar{\phi}(t, x) = \sum_{n \in \mathbb{N}} \phi_{0,n} e^{i\mu_n t} \bar{\theta}_n(x), \tag{8.25}$$

if $\bar{\phi}_0 = \sum_{n \in \mathbb{N}} \phi_{0,n} \bar{\theta}_n$. Once again, $\mu_n = \lambda_n^2$ are the eigenvalues and $\bar{\theta}_n$ are the eigenfunctions of the Dirichlet problem for the laplacian on the network.

The homogeneous system (8.19)-(8.24) is well posed in any of the spaces V^r; the solution is also given by (8.25).

The non-homogeneous system (8.13)-(8.18) is well posed for any $T > 0$, $h \in L^2(0, T)$ and initial datum $\bar{u}_0 \in V'$: there exists a unique solution $\bar{u}$ of (8.13)-(8.18) satisfying $\bar{u} \in C([0, T] : V')$.

For system (8.13)-(8.18) we consider the following control problem: *to determine for which initial data $\bar{u}_0 \in V'$ there exists a function $h \in L^2(0, T)$ such that the solution $\bar{u}$ of (8.13)-(8.18) satisfies $\bar{u}(T, x) = \bar{0}$.* When this is possible, it is said that *the initial datum $\bar{u}_0 \in V'$ is controllable in time T.*

The following proposition provides a characterization of the initial data that are controllable in time T.

Proposition 8.7. *The initial datum $\bar{u}_0 \in V'$ is controllable in time T with control $h \in L^2(0, T)$ if, and only if, for every $\bar{\phi}_0 \in Z$ the following identity is satisfied*

$$i \langle \bar{u}_0, \overline{\bar{\phi}_0} \rangle_H = \int_0^T h(t) \overline{\partial_n \phi^1}(t, \mathbf{v}_1) dt,$$

where $\bar{\phi}$ is the solution of the homogeneous system (8.19)-(8.24) with initial datum $\bar{\phi}_0$.

This characterization may be rewritten as a problem of moments:

Proposition 8.8. *The initial datum $\bar{u}_0 = \sum_{n \in \mathbb{N}} u_{0,n} \bar{\theta}_n \in V'$ is controllable in time T with control $h \in L^2(0, T)$ if, and only if, the following equalities are verified*

$$\int_{-T/2}^{T/2} \varkappa_n e^{-i\mu_n t} h(t) dt = u_{0,n} e^{-i\mu_n T/2}, \qquad n \in \mathbb{N}. \tag{8.26}$$

On the other hand, the HUM allows us giving an alternative characterization:

Proposition 8.9. *The following properties are equivalent:*

a) There exist $T > 0$ and a sequence $(c_n)_{n \in \mathbb{N}}$ of positive numbers such that the inequality

$$\int_0^T \left| \partial_n \phi^1(t, \mathbf{v}_1) \right|^2 dt \geq \sum_{k \in \mathbb{N}} c_k^2 \left| \phi_{0,k} \right|^2, \tag{8.27}$$

is verified by every solution $\bar{\phi}$ of the homogeneous system (8.19)-(8.24) with initial datum $\bar{\phi}_0 \in Z$, or equivalently,

$$\int_0^T \left| \sum_{n \in \mathbb{N}} \varkappa_n a_n e^{i\mu_n t} \right|^2 dt \geq \sum_{n \in \mathbb{N}} c_n^2 \left| a_n \right|^2, \tag{8.28}$$

for every finite sequence (a_n) of complex numbers,

b) the space

$$\mathbb{W} = \left\{ \bar{u}_0 = \sum_{n \in \mathbb{N}} u_{0,n} \bar{\theta}_n \in V' : \sum_{n \in \mathbb{N}} \frac{1}{c_n^2} \left| u_{0,n} \right|^2 < \infty \right\}$$

is controllable in time T.

Theorem 8.10. *System (8.13)-(8.18) is spectrally controllable in any time $T > 0$ if, and only if, $\varkappa_n \neq 0$ for every $n \in \mathbb{N}$.*

Proof. The necessity of the condition $\varkappa_n \neq 0$ is immediate: if $\varkappa_n = 0$ for some value of n then the equality of Proposition 8.8 becomes $u_{0,n} = 0$. Consequently, it is not possible to control an initial datum $\bar{u}_0 = \bar{\theta}_n \in Z$ with $u_{0,n} = 1$.

The proof of the sufficiency can be obtained from Proposition 8.8. The key element is provided by Proposition 6.2, which ensures that

$$\lim_{n \to \infty} \frac{\mu_n}{n^2} = \frac{\pi^2}{L^2}.$$

This implies that the sequence (μ_n) satisfies

$$\sum_{n \in \mathbb{N}} \frac{1}{\mu_n} < \infty. \tag{8.29}$$

As it has been pointed out in section 3.3 of Chapter 3, the problem of moments (8.26) has a solution for any finite sequence

$$m_n = \frac{1}{\varkappa_n} u_{0,n} e^{-i\mu_n T/2}, \qquad n \in \mathbb{N}$$

if it is possible to find a biorthogonal sequence to $\left(e^{i\mu_n t}\right)$.

But the property (8.29) guarantees that for every $\tau > 0$ there exists a non-trivial entire function of exponential type at most τ, vanishing at every μ_n (see, e.g., Theorem 15, p. 139 in [116]). Then, for every $\tau > 0$, there exists a sequence biorthogonal to $\left(e^{i\mu_n t}\right)$ in $L^2(-\tau, \tau)$.

Remark 8.11. It is also possible to give a proof of the sufficiency of the condition $\varkappa_n \neq 0$ for every $n \in \mathbb{N}$ for the spectral controllability of system (8.13)-(8.18) using Proposition 8.9. Indeed, as the sequence (λ_n) has finite upper density $D^+(\lambda_n)$, then $D^+(\mu_n) = 0$. If we apply Corollary 3.33 of Theorem 3.32 it follows that for every $T > 0$ there exist positive numbers γ_n, $n \in \mathbb{N}$, such that

$$\int_0^T \left| \sum_{n \in \mathbb{N}} \varkappa_n a_n e^{i\mu_n t} \right|^2 dt \geq \sum_{n \in \mathbb{N}} \varkappa_n^2 \gamma_n^2 |a_n|^2 ,$$

for any finite sequence (a_n) of complex numbers. This is the inequality (8.27) with $c_n = \varkappa_n \gamma_n$; all these coefficients are positive if $\varkappa_n \neq 0$ for every $n \in \mathbb{N}$.

Corollary 8.12. *For every $T > 0$ the following properties of the system (8.13)-(8.18) are equivalent:*

– the unique continuation of the solutions of the homogeneous system from the controlled node:

$$\partial_n \phi^1(., \mathbf{v_1}) = 0 \quad in \ L^2(0;T) \Rightarrow \bar{\phi}_0 = \bar{0};$$

– the spectral unique continuation from the controlled node:

$$\varkappa_n \neq 0 \quad for \ every \ n \in \mathbb{N}.$$

As for the heat equation, it is possible to describe subspaces of controllable initial data for system (8.13)-(8.18) based on similar informations on the corresponding wave equation.

Proposition 8.13. *If the subspace $\mathcal{W}^r$ is controllable for system (2.11)-(2.16) in some time $T > 0$ for some finite r, then all the initial data $\bar{u}_0 \in V^{2r-1}$ are controllable in any time $\tau > 0$ for system (8.13)-(8.18).*

Proof. According to Remark 3.17, if the subspace $\mathcal{W}^r$ is controllable for system (2.11)-(2.16) in time $T > 0$, then the inequality

$$\int_0^T \left| \sum_{n \in \mathbb{N}} \varkappa_n a_n e^{i\lambda_n t} \right|^2 dt \geq C \sum_{n \in \mathbb{N}} \lambda_n^{2(1-r)} |a_n|^2 , \tag{8.30}$$

is valid for any finite complex sequence (a_n).

Let us observe that if $\mathcal{W}^r$ is controllable in time T for system (2.11)-(2.16) then it is also spectrally controllable and thus $\varkappa_n \neq 0$, $n \in \mathbb{N}$. Besides, proceeding as in Remark 4.31, from (8.30) it follows that there exist constants $C_1, C_2 > 0$ such that for every $n \in \mathbb{N}$,

$$C_1 \lambda_n^{1-r} \leq |\varkappa_n| \leq C_2 \lambda_n. \tag{8.31}$$

Then the inequality (8.30) can be written in the equivalent form

$$\int_0^T \left| \sum_{n \in \mathbb{N}} a_n e^{i\lambda_n t} \right|^2 dt \geq C \sum_{n \in \mathbb{N}} \lambda_n^{2(1-r)} |a_n|^2 |\varkappa_n|^{-2},$$

and from (8.31) we get

$$\int_0^T \left| \sum_{n \in \mathbb{N}} a_n e^{i\lambda_n t} \right|^2 dt \geq C \sum_{n \in \mathbb{N}} \lambda_n^{-2r} |a_n|^2. \tag{8.32}$$

If we apply Theorem 3.34 to the inequality (8.32) we obtain

$$\int_0^T \left| \sum_{n \in \mathbb{N}} a_n e^{i\mu_n t} \right|^2 dt \geq C \sum_{n \in \mathbb{N}} \lambda_n^{-2r} |a_n|^2. \tag{8.33}$$

In view of (8.31), from (8.33) it follows

$$\int_0^T \left| \sum_{n \in \mathbb{N}} \varkappa_n a_n e^{i\mu_n t} \right|^2 dt \geq C \sum_{n \in \mathbb{N}} \lambda_n^{-2r} |a_n|^2 |\varkappa_n|^2 \geq C \sum_{n \in \mathbb{N}} \lambda_n^{2(1-2r)} |a_n|^2. \tag{8.34}$$

Now it suffices to note that, according to Proposition 8.9, the fact that the inequality (8.34) is valid for any finite complex sequence (a_n) is equivalent to the fact that all the initial data $\bar{u}_0 \in V^{2r-1}$ are controllable in time $\tau > 0$ for system (8.13)-(8.18).

Theorem 8.10 and Proposition 8.13 allow obtaining immediate information for the Schrödinger equation on the star-shaped networks studied in section 5.8 of Chapter 5, using the corollaries 5.38, 5.39 and 5.40.

Corollary 8.14. *If the lengths $\ell_1, ..., \ell_{n-1}$ of the uncontrolled edges of a star-shaped network are such that*

1) *all ratios ℓ_i/ℓ_j with $i \neq j$ are irrational numbers, then system (8.13)-(8.18) is spectrally controllable in any time $T > 0$.*
2) *all the ratios ℓ_i/ℓ_j with $i \neq j$ belong to some set $\mathbf{B}_\varepsilon$, then the subspace $V^{2n-4+\varepsilon}$ of initial data for system (8.13)-(8.18) is controllable in any time $T > 0$.*
3) *verify the conditions (S), then for every $\varepsilon > 0$ the subspace $V^{1+\varepsilon}$ of initial data for system (8.13)-(8.18) is controllable in any time $T > 0$.*

Remark 8.15. All the results of this section are valid for the system obtained by replacing equation (8.13) by $iu_t^k + u_{xx}^k = 0$. In this case, the corresponding observability inequality is

$$\int_0^T \left| \sum_{n\in\mathbb{N}} \varkappa_n a_n e^{-i\mu_n t} \right|^2 dt \geq \sum_{n\in\mathbb{N}} c_n^2 |a_n|^2,$$

which, clearly, coincides with (8.28).

Remark 8.16. In these two sections we have derived observability and controllability properties for Schrödinger and heat equations on networks from those we previously obtained on the corresponding wave model. Our tools are based on the moment problem formulation and the Fourier series development of solutions. We also refer to [46] for an application of this method to the study of the controllability of the heat equation in $1 - d$ with low regularity coefficients. However, there are integral transforms allowing to do the same kind of transfer of results from one model to the others that work also for equations in domains of $\mathbb{R}^n$ in any dimension (see, for instance, [87], [88], [86], and [98]).

8.3 A Model of Network for Beams

Now we consider the following model of a network of flexible beams controlled from one exterior node.

$$u_{tt}^i + u_{xxxx}^i = 0 \qquad\qquad \text{in } \mathbb{R} \times [0, \ell_i], \quad i = 1, ..., M,$$
$$\tag{8.35}$$

$$u^1(t, \mathbf{v}_1) = 0, \qquad \partial_n^2 u^1(t, \mathbf{v}_1) = h(t) \qquad t \in \mathbb{R}, \tag{8.36}$$
$$u^{i(j)}(t, \mathbf{v}_j) = 0 \qquad\qquad t \in \mathbb{R}, \quad j = 2, ..., N, \tag{8.37}$$

$$u^i(t, \mathbf{v}) = u^j(t, \mathbf{v}), \qquad \partial_n^2 u^i(t, \mathbf{v}) = \partial_n^2 u^j(t, \mathbf{v}) \quad t \in \mathbb{R}, \quad \mathbf{v} \in \mathcal{V}_\mathcal{M}, \, i, j \in I_\mathbf{v},$$
$$\tag{8.38}$$

$$\sum_{i\in I_\mathbf{v}} \partial_n u^i(t, \mathbf{v}) = \sum_{i\in I_\mathbf{v}} \partial_n^3 u^i(t, \mathbf{v}) = 0 \qquad t \in \mathbb{R}, \quad \mathbf{v} \in \mathcal{V}_\mathcal{M}, \tag{8.39}$$

$$u^i(0, x) = u_0^i(x), \qquad u_t^i(0, x) = u_1^i(x) \qquad x \in [0, \ell_i], \quad i = 1, ..., M.$$
$$\tag{8.40}$$

Let us observe that in this case the control acts through the second normal derivative $\partial_n^2 u^1(., \mathbf{v}_1)$ at the node $\mathbf{v}_1$.

System (8.35)-(8.40) is well posed for $h \in L^2(0, T)$ and $\bar{u}_0 \in V$, $\bar{u}_1 \in V'$. The homogeneous version of (8.35)-(8.40) is also well posed for $\bar{u}_0 \in V^2$, $\bar{u}_1 \in H$.

We study the following control problem in time T for system (8.35)-(8.40): to determine for which initial states $(\bar{u}_0, \bar{u}_1) \in V \times V'$, there exists $h \in L^2(0, T)$ such that the corresponding solution $\bar{u}$ of (8.35)-(8.40) satisfies

$$\bar{u}(T, .) = \bar{u}_t(T, .) = \bar{0}.$$

Those initial states $(\bar{u}_0, \bar{u}_1)$ for which such a function h exists will be said to be *controllable in time T*. We will say that a subspace of $V \times V'$ is controllable in time T if so are all of its elements. In particular, if $Z \times Z$ is controllable in time T, we will say that system (8.35)-(8.40) is *spectrally controllable in time T*.

Let us remark that Z denotes, as previously, the space of all the finite linear combinations of the eigenfunctions of the operator $\mathcal{D}_G$ associated to (8.35)-(8.40). This is the operator $\mathcal{D}_G : H \to H$ defined by

$$\mathcal{D}_G(u^1, ..., u^M) = (u^1_{xxxx}, ..., u^M_{xxxx})$$

with the boundary conditions

$$u^{i(\mathbf{v})}(\mathbf{v}) = \partial_n^2 u^{i(\mathbf{v})}(\mathbf{v}) = 0,$$

at the exterior nodes and

$$u^i(\mathbf{v}) = u^j(\mathbf{v}), \qquad \partial_n^2 u^i(\mathbf{v}) = \partial_n^2 u^j(\mathbf{v}) \qquad i, j \in I_{\mathbf{v}},$$

$$\sum_{i \in I_{\mathbf{v}}} \partial_n u^i(\mathbf{v}) = \sum_{i \in I_{\mathbf{v}}} \partial_n^3 u^i(\mathbf{v}) = 0$$

at the interior ones.

The operator $\mathcal{D}_G$ coincides with the square of the elliptic operator $-\Delta_G$ associated to (2.11)-(2.16). By this reason, the eigenfunctions of $\mathcal{D}_G$ coincide with the eigenfunctions $(\bar{\theta}_n)$ of $-\Delta_G$, the eigenvalues being (μ_n^2). In particular, the space Z for equation (8.35)-(8.40) coincides with that of (2.11)-(2.16). Besides, the solution of the homogeneous system (8.35)-(8.40) (with $h \equiv 0$) with initial data

$$\bar{\phi}_0 = \sum_{n \in \mathbb{N}} \phi_{0,n} \bar{\theta}_n, \qquad \bar{\phi}_1 = \sum_{n \in \mathbb{N}} \phi_{1,n} \bar{\theta}_n,$$

is expressed by the formula

$$\bar{\phi}(t, x) = \sum_{n \in \mathbb{N}} \left(\phi_{0,n} \cos \mu_n t + \frac{\phi_{1,n}}{\mu_n} \sin \mu_n t \right) \bar{\theta}_n(x).$$

Proposition 8.17. *The initial state $(\bar{u}_0, \bar{u}_1) \in V \times V'$ is controllable in time T with control $h \in L^2(0, T)$ if, and only if, for every $(\bar{\phi}_0, \bar{\phi}_1) \in Z \times Z$ the following equality is true*

$$\langle \bar{\phi}_1, \bar{u}_0 \rangle_{V' \times V} - \langle \bar{u}_1, \bar{\phi}_0 \rangle_{V' \times V} = \int_0^T h(t) \partial_n \phi^1(t, \mathbf{v}_1) dt, \qquad (8.41)$$

where $\bar{\phi}$ is the solution of the homogeneous system (8.35)-(8.40) with initial state $(\bar{\phi}_0, \bar{\phi}_1)$.

Clearly, it is sufficient to check (8.41) for the initial states of the form $(\bar{0}, \bar{\theta}_n)$ and $(\bar{\theta}_n, \bar{0})$, $n \in \mathbb{N}$. Then, if we define $\mu_n = -\mu_{-n}$ for $n < 0$, Proposition 8.17 gives rise to a problem of moments:

Proposition 8.18. *The initial state $(\bar{u}_0, \bar{u}_1) \in V \times V'$ is controllable in time T with control $h \in L^2(0, T)$ if, and only, the equalities*

$$\int_0^T \varkappa_{|n|} h(t) e^{i\mu_n t} dt = u_{1,|n|} - i\mu_n u_{0,|n|}, \qquad (8.42)$$

are verified for every $n \in \mathbb{Z}_*$.

Let us observe that the problem of moments (8.42) coincides with that of the Schrödinger equation, except by the fact that now the sequence $(\mu_n)_{n \in \mathbb{N}}$ should be replaced by $(\mu_n)_{n \in \mathbb{Z}_*} = (\pm\mu_n)_{n \in \mathbb{N}}$. That is why, proceeding as in the proof of Theorem 8.10 it is possible to prove:

Theorem 8.19. *The system (8.35)-(8.40) is spectrally controllable in any time $T > 0$ if, and only if, $\varkappa_n \neq 0$ for every $n \in \mathbb{N}$.*

On the other hand, the HUM allows us to obtain from Proposition 8.17:

Proposition 8.20. *The following two facts are equivalent:*
a) There exist $T > 0$ and a sequence $(c_n)_{n \in \mathbb{N}}$ of positive numbers such that the following inequality is verified

$$\int_0^T |\partial_n \phi^1(t, \mathbf{v}_1)|^2 dt \geq \sum_{k \in \mathbb{N}} c_k^2 \left(\mu_n^2 \phi_{0,k}^2 + \phi_{1,k}^2 \right), \qquad (8.43)$$

by every solution $\bar{\phi}$ of the homogeneous system (8.35)-(8.40) with initial state $(\bar{\phi}_0, \bar{\phi}_1) \in Z \times Z$, or equivalently, the inequality

$$\int_0^T \left| \sum_{n \in \mathbb{Z}_*} \varkappa_{|n|} a_n e^{i\mu_n t} \right|^2 dt \geq \sum_{n \in \mathbb{N}} c_n^2 |a_n|^2, \qquad (8.44)$$

is valid for every finite sequence of complex numbers (a_n) verifying $a_{-n} = \overline{a_n}$.
b) The space

$$W = \left\{ (\bar{u}_0, \bar{u}_1) \in V \times V' : \sum_{n \in \mathbb{N}} \left(\frac{1}{c_n^2} u_{0,n}^2 + \frac{1}{c_n^2 \mu_n^2} u_{1,n}^2 \right) < \infty \right\}$$

is controllable in time T.

Once again, it is possible to identify subspaces of controllable initial states for system (8.35)-(8.40) based on the similar information for (2.11)-(2.16).

Proposition 8.21. *If the subspace $\mathcal{W}^r$ is controllable for (2.11)-(2.16) in time $T > 0$, then all the initial states $(\bar{u}_0, \bar{u}_1) \in V^{2r+1} \times V^{2r-1}$ are controllable in any time $\tau > 0$ for (8.35)-(8.40).*

As a consequence of the previous result, for star-shaped networks the following holds:

Corollary 8.22. *If the lengths $\ell_1, ..., \ell_{n-1}$ of the uncontrolled edges of a star-shaped network are such that*

1) *all the ratios ℓ_i/ℓ_j with $i \neq j$ are irrational numbers, then system (8.35)-(8.40) is spectrally controllable in any time $T > 0$.*
2) *all the ratios ℓ_i/ℓ_j with $i \neq j$ belong to some set $\mathbf{B}_\varepsilon$, then the subspace $V^{2n-2+\varepsilon} \times V^{2n-4+\varepsilon}$ of initial states for system (8.35)-(8.40) is controllable in any time $T > 0$.*
3) *verify the conditions (S), then for every $\varepsilon > 0$ the subspace $V^{3+\varepsilon} \times V^{1+\varepsilon}$ of initial states for system (8.35)-(8.40) is controllable in any time $T > 0$.*

9

Final Remarks and Open Problems

9.1 Brief Description of the Main Results of the Book

9.1.1 Networks of Strings

The main result on the spectral controllability of arbitrary networks from an exterior node is given in Theorem 6.5: *the network is spectrally controllable in some finite time if and only if the spectral unique continuation property from the controlled node is verified. Besides, when the spectral unique continuation property holds, the network is spectrally controllable in any time larger than twice the total length of the network; which turns out to be the minimal spectral control time.* From this point of view, the networks of strings behave, essentially, as a single string whose length coincides with the total length of the network controlled on an interior point. The main reason is that the sequence of eigenvalues of the network is asymptotically equivalent to the sequence of eigenvalues of a string with that length (Proposition 6.2).

The main difference between those cases consists in that, for a single string the spectral unique continuation property is verified under an irrationality condition, while for any non-trivial topological configuration of the network we do not have a simple condition for the spectral unique continuation property to be fulfilled. Moreover, except for the simplest case in which the network is reduced to a single string, the exact controllability of networks is never reached by means of one single control (Theorem 2.8). In this sense, the exterior control of a network from one node is analogous to the control of a string from an interior point.

The spectral controllability property allows to ensure the controllability of a subspace of initial states that may be explicitly described in terms of the eigenfunctions of the network and the traces of the normal derivatives of the eigenfunctions at the controlled node (Remark 6.6).

For networks with simple topological configurations it is possible to provide more precise information:

Tree-shaped networks. When the network graph is a tree it is possible to obtain a complete characterization of those trees for which the spectral unique continuation property is verified (Proposition 5.26) and thus, to characterize the trees which are spectrally controllable in a time equal to twice the total length of the network. The set of trees with a given topological configuration for which the spectral unique continuation fails has null Lebesgue measure (Proposition 5.32). Though these results could be obtained from Theorem 6.5, the propagation technique used in Chapter 5, essentially based on the representation of the solutions of the wave equation by means of the d'Alembert formula, allows to prove the spectral controllability in the minimal time (Theorem 5.23). Besides, it provides a weighted observability inequality with weights that are explicitly computed in terms of the eigenvalues.

Some of the results are of independent interest. That is the case, for instance, of the compatibility conditions $\mathcal{P}u_t(.,\mathbf{v}) + \mathcal{Q}u_x(.,\mathbf{v}) = 0$ at the controlled node (Proposition 5.6). From those conditions it is possible to obtain an equation for the eigenvalues (Proposition 5.19) and the pseudo-periodicity property of the solutions of the homogeneous system (Remark 5.12), which implies that increasing the control time does not lead to improving the controllability results (Proposition 5.4).

Star-shaped networks.   The star-shaped networks are a particular case of trees. In this case, the spectral unique continuation condition means that the ratios of the lengths of the uncontrolled strings are irrational numbers (Section 5.8.1). Besides, it is possible to identify subspaces of controllable initial states of the form $\mathcal{W}^r$ (which are, essentially, Sobolev spaces on the strings with appropriate boundary and coupling conditions at the multiple node). The structure of such subspaces depends on rational approximation properties of the ratios of the lengths of the strings in the network.

For these networks it is possible to prove that when the observation time is smaller than twice the total length of the network, the unique continuation property from the controlled node fails for the solutions of the corresponding homogeneous system. This implies not only the lack of spectral controllability, but also of the approximate one. In Section 4.9 we have given an example of a smooth solution of the homogeneous system for which the unique continuation property is not true when the observation time is small.

9.1.2 Simultaneous Control of Strings

Simultaneous control of trees from one exterior node. The results concerning the simultaneous control from an exterior node of a finite number of tree-shaped networks (Corollary 5.37) are similar to those corresponding to a single tree: *the networks are simultaneously spectrally controllable in some finite time, if and only if each of them is spectrally controllable and the spectra of the networks are pairwise disjoint. The minimal time to simultaneously control the system is the sum of the minimal control times of all the networks.*

When the networks are simultaneously spectrally controllable it is possible to indicate subspaces of controllable initial states, which are explicitly defined in terms of the eigenvalues of the networks (Proposition 5.36). In particular, if the simultaneously controlled networks are strings, then for certain values of the lengths of the strings it is possible to build Sobolev-type spaces of controllable initial states (Corollaries 5.42 and 5.43). Besides, it is proved (Corollary 4.12) that, depending on the lengths of the strings, the space of controllable initial states may be arbitrarily small, that is, there exist initial states with Fourier coefficients increasing arbitrarily rapidly, which are not controllable in any finite time.

Control at all the nodes with only four control functions. Using the results on the simultaneous control of strings, one can determine how many different controls are needed to reach the spectral controllability of the network. A simple application of the Four Colors Theorem allows to ensure that, under certain irrationality conditions on the lengths of the strings, four colors are sufficient (besides, one of them may be chosen to be identically equal to zero) to control the network in a time T^* (and then in any time larger than T^*), which is smaller than twice the total length of the network (Proposition 6.8).

Simultaneous interior control of strings. The simultaneous control from an open set of two strings with different densities turns out to be a much more robust property: simultaneous exact controllability holds in any time larger than the characteristic times of both strings (Corollary 7.2). However, when the strings are of the same density, the results are completely analogous to those obtained for the simultaneous control of two strings from one extreme (Corollaries 7.10 and 7.11).

9.1.3 Other Equations on Networks

For Shrödinger (Theorem 8.10), heat (Theorem 8.3) and beam (Theorem 8.19) equations the spectral unique continuation property from the controlled node is necessary and sufficient for the system to be spectrally controllable in any arbitrarily small time.

In those cases in which spaces of the type W^r of controllable initial states are known for the wave equation (which is the case, for instance, of the star-shaped networks), then it is possible to identify subspaces of controllable initial data for the Schrödinger, heat and beam equations on that network (Propositions 8.5, 8.13, 8.21, respectively). In particular, the heat equation is null-controllable or controllable to zero.

9.2 Future Lines of Research and Open Problems

The field of multi-structures is extremely rich and provides a large number of interesting problems of quite complex mathematical nature. The mathematical study of these problems will necessarily require of/lead to new, more powerful mathematical tools.

In our opinion, the future development in connection with the problems addressed in this book should follow, at least, the following main directions:

1.– Study of more complex models of string and beam networks, providing a more realistic description of the motion of these objects and, in particular, taking into account their three-dimensional character. The book [68] provides a valuable source of models of multi-body structures. We also refer to the article [63] for an account of the main developments in this field.

2.– Equations with variable coefficients on graphs. Some of the techniques developed in Chapters 6 and 8 might be of use to deal with this more complex situations, but much remains to be done for a complete understanding of the issue. In particular, important further work is needed to address equations with coefficients depending both on space and time, and with low regularity.

3.– Nonlinear problems on $1-d$ networks. Most of the existing results on the controllability of semilinear PDE in open domains of $\mathbb{R}^n$ rely on fixed point arguments and a sharp analysis of the cost of control and/or observability for linearized equations with potentials. Taking into account that the latter is mainly open, the controllability of semilinear models on networks is it as well.

4.– Stabilization problems. Controllability and stabilization problems are closely related. This monograph has been fully devoted to the problems of control. One may expect that each of the controllability results presented here, mostly for wave-like networks, will have some counterparts in the context of stabilization too. But this issue is widely open.

5.– Coupled systems of multi-dimensional flexible objects. A complete analysis of this issue will require the use of microlocal tools. As mentioned above, the analysis in [25] on the polarization of singularities may be particularly relevant in this context. The recent spectral characterization of the controllability of a general abstract wave-like system in [101] may also be helpful in this context. On the other hand, solutions of PDE in multi-dimensional networks often develop singularities on the interfaces. The topic of the controllability of wave equations in the presence of singularities started with the work by P. Grisvard [47], and some extensions to the case of networks of $2-d$ domains were developed by S. Nicaise [92] and [93]. But a systematic study of these issues is still to be done.

6.– Carleman inequalities on networks and related issues. One of the most flexible tools to deal with observability problems for PDE in domains of the euclidean space are the Carleman inequalities. They allow handeling both hyperbolic and parabolic models, as well as Schrödinger and plate equations. This subject is mainly open in the context of networks. We refer to [41] for the analysis of a model coupling two $1 - d$ heat equations by means of a point mass, using Carleman inequalities.

7.- Numerical approximation issues. Recently, important progresses have been done on this topic and in particular on clarifying whether controlling a numerical approximation model is an efficient way to get convergent approximations of controls ([124]). Very little is known in the context of networks.

As we said above, a lot remains to be done for a complete theoretical understanding of these issues.

To be more precise, let us mention some open problems which are directly related to the content of this book.

1.- Observability properties of wave equations with potentials $\alpha = \alpha(x, t)$ of the form

$$u_{tt} - u_{xx} - \alpha(x, t)u = 0$$

on graphs are mainly unknown.

This problem has been recently considered in [60] in the case of the simultaneous observation of strings for constant α. In that paper, using the generalized Ingham Theorem 3.29, it is proved that, generically in the set of all possible lengths of the strings, the unique continuation property and the spectral controllability hold. However, it would be interesting to describe the observed norm in terms of Sobolev spaces as it has been done for the case $\alpha = 0$ in this book. The case of variable α is a completely open question. Note that, for the scalar wave equation, one can treat this case as a perturbation of the wave equation without potential, because of the compactness of the zero order perturbation in the energy space. But this argument fails in the case of networks, because of the intrinsic loss of regularity of solutions in the observation process. When $\alpha = \alpha(x)$ the Fourier series arguments developed in chapter 6 should be sufficient to analyze spectral controllability. On the other hand, the analogue of Schmidt's result in section 2.4 (guaranteeing that a tree-like network can be exactly controlled by means of controls acting on all except for one external node), based on sidewise energy estimates, can be easily extended to this case even if α depends both on x and t. But a systematic analysis of the problem with potential depending both in x and t and in general networks is far from being complete.

2.- The same problems can be formulated for more general wave equations with variable coefficients like, for instance,

$$\rho(x)u_{tt} - (\sigma(x)u_x)_x = 0.$$

The general results derived as a consequence of the application of Beurling-Malliavin's Theorem in section 6.1 are expected to be still true in this more general case. It would be sufficient to analyze the spectral density, but the same arguments we have employed there based on the comparison of the spectrum with the Dirichlet and Neumann ones on individual strings should apply. But all the other results, based on propagation arguments, will need important further developments to be applied in this context. One can for instance try to use the classical trick of performing a change of variable to reduce the problem to an equation of the form

$$v_{tt} - v_{xx} + a(x)v = 0.$$

But then, we return to the difficulty mentioned in problem # 1.

Note also that the methods in section 6.1 using spectral arguments fail if the coefficients depend both on x and t.

3.– The controllability of the semilinear equation

$$u_{tt} - u_{xx} + f(u) = 0$$

in networks does not seem to be addressable with the tools developed in this book. Indeed, the existing techniques for treating nonlinear problems rely on linearization arguments and fixed point techniques. Therefore, the problem # 1 of the wave equation with potential seems to be a preliminary step to address semilinear equations. Once more, the particular case of the tree-like network with control in all except for one external node could be handeled by the existing techniques, using Schmidt's Theorem in section 2.4. We refer to [119] for the proof of controllability in the case of one single string and to [117] for the state of the art in the context of the control of the multi-dimensional wave equation.

Of course, similar questions arise for more general systems of nonlinear hyperbolic equations. We refer to [48] and [74] for the analysis of the controlability of St. Venant equations in networks.

4.– The result of section 6.1 characterizes the property of spectral controllability of a general graph in terms of a natural condition on the spectrum: Every eigenfunction should have a non-trivial normal trace on the controlled node. In the case of trees we were able to give a further characterization of this condition: the spectra of all sub-trees meeting at some node should have empty intersection. It would be interesting to generalize this condition to general networks. The main difficulty for doing this is the presence of closed loops or cycles on the network. The problem of characterizing the space of controllable data is open even for the simplest networks with cycles.

5.– A quite rich and widely open field of research is that of the numerical computation of controls for networks of strings in the spirit of [53] and [122]. We refer to [124] for a description of the state of the art in the case of one single equation, mainly in one space dimension. It is known that controlling a numerical approximation scheme is not sufficient to guarantee the convergence of controls. Rather, the numerical approximation scheme has to be chosen with appropriate dispersive properties so that uniform (with respect to the mesh size) bounds on the control and their convergence holds. Two-grid algorithms are also useful to guarantee convergence (see [89]). But very little is known for networks, except for the work [22] in which an extension of the results in [53] to the case of a tripod of strings is given. In particular, the possibility of using the two-grid algorithm on networks to recover the positive results on spectral controllability in this book is open.

The convergence of domain decomposition techniques has been also analyzed for optimal control problems. We refer to the book [66] for a complete account of the state of the art in this context.

6.– The complexity of the problems under consideration also increases when the flexible structure under consideration connects objects of different dimensions. This subject is widely open. We refer to [49] (resp. [28]) for the analysis of a simple model connecting two strings (resp. beams) with a point mass and to [38] for a nonlinear version of this kind of models. In [100] and [24] a model for the coupling between a $2 - d$ and a $1 - d$ wave equation is also analyzed. We also refer to [99] for the analysis of the same issue for beam and Schrödinger equations.

7.– As we mentioned above, the problems of controllability and stabilization are closely related ([59], [121] and [123]). In stabilization one aims to ensure the decay of the energy of solutions by means of a feedback control. Very likely, each of the results of this monograph may lead to its counterpart in the context of stabilization. But this subject is widely open. Note that, in general, when exact controllability holds, one expects an uniform exponential decay rate of the energy of solutions. However, when controllability holds in smaller subspaces (than the natural and optimal energy space) one may only expect slower decay rates for smooth solutions (see [57]). We refer to [68] and [7] for partial results in this direction.

8.– In this monograph we have considered only networks in which all equations have the same nature. However, in practice and, in particular, in fluid-structure interaction, models in which heat and wave-like equations are coupled arise naturally. This problem has been recently investigateed in a number of articles in a domain of the $\mathbb{R}^n$ (see for instance [118] for the $1 - d$ case) . This subject is completely open in the context of networks. The same can be said about the system of thermoelasticity ([120], [71]).

9.– As indicated in section 4.10 the moment problem techniques need to be further developed in order to give the sharp controllability results we have obtained by means of propagation techniques.

10.– As we said in the introduction, the problem of observation of solutions on networks from its boundary is also closely related to the theory of inverse problems. This subject is also mainly open. We refer to [95] for some results in this frame.

A

Some Consequences of Diophantine Approximation Theorems

In this appendix we have gathered some results that have been used in the proof of several theorems in the main text. All of these results have a common feature: they are obtained as consequences of classical theorems on the approximation of real numbers by rational ones.

For $\eta \in \mathbb{R}$ we denote by $|||\eta|||$ the distance from η to the set $\mathbb{Z}$:

$$|||\eta||| := \min_{\eta - x \in \mathbb{Z}} |x|$$

and by $\mathbf{E}(\eta)$ the closest integer number [1] to η:

$$|\eta - \mathbf{E}(\eta)| = |||\eta|||.$$

Let us observe that $0 \leq |||\eta||| \leq 1/2$ and besides, that η may be expressed as

$$\eta = \mathbf{E}(\eta) + \mathbf{F}(\eta), \tag{A.1}$$

where

$$|\mathbf{F}(\eta)| = |||\eta|||, \qquad -\frac{1}{2} \leq \mathbf{F}(\eta) \leq \frac{1}{2}.$$

For given real numbers $\ell_1, ..., \ell_N$, we define the function

$$\mathbf{a}(\lambda) = \mathbf{a}(\lambda, \ell_1, ..., \ell_N) := \sum_{i=1}^{N} \prod_{j \neq i} |\sin \lambda \ell_j|.$$

This function frequently appears in the problems we have studied. Our aim is to find conditions on the numbers $\ell_1, ..., \ell_N$ guaranteeing that for some $\alpha \in \mathbb{R}$ the function

$$\mathbf{a}(\lambda)\lambda^\alpha$$

---

[1] If $|||\eta||| = 1/2$ there are two integer numbers with that property: $\eta + 1/2$ and $\eta - 1/2$. In this case $\mathbf{E}(\eta)$ will denote one of these numbers.

remains bounded from below as $\lambda \to \infty$.

For $m \in \mathbb{N}$, $i = 1, ..., N$, we denote

$$\mathbf{z}^i(m) := \mathbf{z}^i(m, \ell_1, ..., \ell_N) := \prod_{j \neq i} |||\frac{\ell_j}{\ell_i} m|||,$$

$$\mathbf{m}^i(\lambda) := \mathbf{m}^i(\lambda, \ell_1, ..., \ell_N) := \mathbf{E}\left(\frac{\ell_i}{\pi}\lambda\right).$$

The following proposition allows to reduce the stated problem to one on the approximation by rational ones.

Proposition A.1. *There exists a positive constant C such that for every $\lambda \in \mathbb{R}$ the inequality*

$$\mathbf{a}(\lambda) \geq C \min_{i=1,...,N} \mathbf{z}^i(\mathbf{m}^i(\lambda))$$

is satisfied.

Proof. Let us remark first that every $x \in \mathbb{R}$ can be expressed in the form

$$x = \pi \mathbf{E}\left(\frac{x}{\pi}\right) + \pi \mathbf{F}\left(\frac{x}{\pi}\right).$$

Then it follows

$$\sin x = \sin \pi \mathbf{F}\left(\frac{x}{\pi}\right). \tag{A.2}$$

Taking into account that if $|x| \leq 1/2$ then the inequalities

$$2|x| \leq |\sin \pi x| \leq \pi |x|,$$

are true, from (A.2) we obtain for every $x \in \mathbb{R}$

$$2|||\frac{x}{\pi}||| \leq |\sin x| \leq \pi |||\frac{x}{\pi}|||. \tag{A.3}$$

In view of this inequality, we have for every $\lambda \in \mathbb{R}$

$$2|||\mathbf{m}^i(\lambda)\frac{\ell_j}{\ell_i}||| \leq \sin\left(\mathbf{m}^i(\lambda)\frac{\ell_j}{\ell_i}\pi\right). \tag{A.4}$$

If we denote

$$\gamma^i = \mathbf{F}\left(\frac{\ell_i}{\pi}\lambda\right),$$

then

$$\mathbf{m}^i(\lambda) = \mathbf{E}\left(\frac{\ell_i}{\pi}\lambda\right) = \frac{\ell_i}{\pi}\lambda - \mathbf{F}\left(\frac{\ell_i}{\pi}\lambda\right) = \frac{\ell_i}{\pi}\lambda - \gamma^i.$$

Replacing this expression of $\mathbf{m}^i(\lambda)$ in the right hand term of (A.4) it follows

$$2|||\mathbf{m}^i(\lambda)\frac{\ell_j}{\ell_i}||| \le \left| \sin\left(\ell_j\lambda - \gamma^i\frac{\ell_j}{\ell_i}\pi \right) \right| = \left| \sin\ell_j\lambda\cos\gamma^i\frac{\ell_j}{\ell_i}\pi - \cos\ell_j\lambda\sin\gamma^i\frac{\ell_j}{\ell_i}\pi \right|$$

$$\le |\sin\ell_j\lambda| + |\gamma^i|\frac{\ell_j}{\ell_i}\pi.$$

On the other hand, from the inequality (A.3) we get

$$|\gamma^i| = |||\lambda\frac{\ell_i}{\pi}||| \le \frac{1}{2}|\sin\lambda\ell_i|.$$

Thus we may conclude that for every $i = 1, ..., N$ and every $j \ne i$

$$|||\mathbf{m}^i(\lambda)\frac{\ell_j}{\ell_i}||| \le \frac{1}{2}|\sin\ell_j\lambda| + \frac{\pi\ell_j}{4\ell_i}|\sin\lambda\ell_i|.$$

Multiplying these inequalities we obtain that there exists a constant $C > 0$ such that for every $i = 1, .., N$

$$\mathbf{z}^i(\mathbf{m}^i(\lambda)) = \prod_{j\ne i}|||\frac{\ell_j}{\ell_i}\mathbf{m}^i(\lambda)||| \le C|\sin\lambda\ell_i| + \frac{1}{2^{N-1}}\prod_{j\ne i}|\sin\ell_j\lambda|. \qquad (A.5)$$

With this, it is now quite simple to prove the assertion of the proposition. We will prove that if (λ_n) is a sequence such that $\mathbf{a}(\lambda_n) \to 0$ then there exists a value i_0 of i such that $\mathbf{z}^{i_0}(\mathbf{m}^{i_0}(\lambda_n)) \to 0$.

Indeed, let us observe that if the sequence (λ_n) satisfies $\mathbf{a}(\lambda_n) \to 0$ then, for every i

$$\prod_{j\ne i}|\sin\ell_j\lambda_n| \to 0.$$

Then there would exist some i_0 such that $|\sin\lambda_n\ell_{i_0}| \to 0$. Thus, from the inequality (A.5) for $i = i_0$ it follows

$$\mathbf{z}^{i_0}(\mathbf{m}^{i_0}(\lambda_n)) \to 0.$$

This proves the proposition.

Corollary A.2. *Let $\alpha \in \mathbb{R}$. If for every $i = 1, ..., N$, the ratio ℓ_j/ℓ_i, $j = 1, ..., N$, has the property that there exists a constant $C_i > 0$ such that, for every $m \in \check{\mathbb{N}}$,*

$$m^\alpha\prod_{j\ne i}|||\frac{\ell_j}{\ell_i}m||| \ge C_i, \qquad (A.6)$$

then

$$\mathbf{a}(\lambda)\lambda^\alpha \ge C,$$

for every $\lambda > 0$.

In what follows we will see some rational approximation theorems which provide sufficient conditions for an inequality like (A.6) to be true.

Let us recall that a real number ξ is said to be algebraic if ξ is the root of some polynomial with rational coefficients. The set $\mathbb{A}$ of all the algebraic numbers is a sub-field of $\mathbb{R}$. The Lebesgue measure of $\mathbb{A}$ is equal to zero, but $\mathbb{A}$ is dense in $\mathbb{R}$. Besides, so is $\mathbb{A} \setminus \mathbb{Q}$. It is said that the algebraic number ξ is of order p if the polynomial with rational coefficients of minimal degree that vanishes at ξ is of degree p.

A classical problem in Number Theory is the following: given $\xi, \alpha \in \mathbb{R}$, to determine whether the inequality

$$|||\xi m||| \leq \frac{1}{m^\alpha}$$

has solutions $m \in \mathbb{Z}$. This is equivalent to the existence of $m, n \in \mathbb{Z}$ such that

$$\left| \xi - \frac{n}{m} \right| \leq \frac{1}{m^{\alpha+1}}.$$

The relevance of this problem is related to the following theorem due to Liouville:

Theorem A.3 (Liouville). *If ξ is an algebraic number of order $p \geq 2$, then the inequality*

$$\left| \xi - \frac{n}{m} \right| \leq \frac{1}{m^p}.$$

has no solutions $n, m \in \mathbb{Z}$.

This fact was used by Liouville to prove that not all the real numbers are algebraic; he constructed explicit examples of numbers $\xi \in \mathbb{R}$, known now as Liouville's numbers, one of which is

$$\xi = \sum_{n=1}^{\infty} 10^{-n^{n^2}},$$

such that the inequality (A.6) has solutions for every p. Clearly, in view of Theorem A.3, such numbers are not algebraic. Nowadays, this fact may seem to be rather simple, since the set of algebraic numbers has Lebesgue measure equal to zero and thus, most real numbers are not algebraic. However, at its time, this result had an outstanding scientific relevance.

Later on, Roth proved a stronger result:

Theorem A.4 (Roth, [104]). *If ξ is an algebraic number of order $p \geq 2$, then the inequality*

$$\left| \xi - \frac{n}{m} \right| \leq \frac{1}{m^2} \tag{A.7}$$

has at most a finite number of solutions $n, m \in \mathbb{Z}$.

The following results provide additional information

Proposition A.5 ([26], p. 120). *For every $\varepsilon > 0$ there exists a set $\mathbf{B}_\varepsilon \subset \mathbb{R}$, such that the Lebesgue measure of $\mathbb{R} \setminus \mathbf{B}_\varepsilon$ is equal to zero, and a constant $C_\varepsilon > 0$ for which, if $\xi \in \mathbf{B}_\varepsilon$ then*

$$|||\xi m||| \geq \frac{C_\varepsilon}{m^{1+\varepsilon}}.$$

On the other hand, when $\varepsilon = 0$ a complete answer can be given. Let $\mathcal{F}$ be the set of all those irrational numbers $\eta \in \mathbb{R}$ such that if $[a_0, a_1, ..., a_n, ...]$ is the expansion of η in continuous fraction (see, e.g., [26] for a definition) then the sequence (a_n) is bounded. The set $\mathcal{F}$ is not denumerable and has Lebesgue measure equal to zero.

Proposition A.6 ([69], Theorem 6, p. 24). *There exists a positive constant C such that*

$$|||\xi m||| \geq \frac{C}{m},$$

for every $m \in \mathbb{N}$, if, and only if, $\xi \in \mathcal{F}$.

In particular, $\mathcal{F}$ is contained in the sets $\mathbf{B}_\varepsilon$ for every $\varepsilon > 0$.

The following theorem due to W. Schmidt provides information on the simultaneous approximation of real numbers by rational ones with the same denominator n.

Theorem A.7 (W. Schmidt, [109]). *If the numbers $\xi_1, ..., \xi_N$ are algebraic and $1, \xi_1, ..., \xi_N$ are linearly independent over the field $\mathbb{Q}$, for every $\varepsilon > 0$, the inequality*

$$|||n\xi_1||| \cdot |||n\xi_2||| \cdots |||n\xi_N|||n^{1+\varepsilon} \leq 1$$

has at most a finite number of solutions $n \in \mathbb{N}$.

An immediate consequence of this theorem is that, if the numbers $\xi_1, ..., \xi_N$ are algebraic and $1, \xi_1, ..., \xi_N$ are $\mathbb{Q}$-linearly independent then, for each $\varepsilon > 0$ there exists a constant $C_\varepsilon > 0$ such that

$$|||n\xi_1||| \cdot |||n\xi_2||| \cdots |||n\xi_N|||n^{1+\varepsilon} \geq C_\varepsilon,$$

for every $n \in \mathbb{N}$.

As a counterpart, Schmidt proved the following more exact version of the Dirichlet theorem.

Theorem A.8 (W. Schmidt, [109]). *If $\xi^1, ..., \xi^M$ are real numbers and $\varepsilon < 1/M$ then, for an infinite number of values of $p \in \mathbb{Z}$, there exist integer numbers $q_i(n)$, $i = 1, ..., M$ such that*

$$|p\xi_i - q_i(p)| \leq \frac{1}{p^\eta} \qquad i = 1, ..., M. \tag{A.8}$$

Definition A.9. *We will say that the real numbers $\ell_1, ..., \ell_N$ verify the conditions (S) if*

- $\ell_1, ..., \ell_N$ *are linearly independent over the field $\mathbb{Q}$ of rational numbers;*
- *the ratios ℓ_i/ℓ_j are algebraic numbers for $i, j = 1, ..., N$.*

Let us observe that if $\ell_1, ..., \ell_N$ verify the conditions (S), then for every i the ratios ℓ_j/ℓ_i, $j = 1, ..., N$, satisfy the conditions of the Schmidt's theorem. Actually, if ℓ_i and ℓ_j are algebraic numbers, so is their ratio. Besides, if

$$\alpha_1 \frac{\ell_1}{\ell_i} + \cdots + \alpha_{i-1} \frac{\ell_{i-1}}{\ell_i} + \alpha_i \cdot 1 + \alpha_{i+1} \frac{\ell_{i+1}}{\ell_i} + \cdots + \alpha_N \frac{\ell_N}{\ell_i} = 0, \qquad \alpha_i \in \mathbb{Q},$$

then

$$\alpha_1 \ell_1 + \cdots + \alpha_N \ell_N = 0, \qquad \alpha_i \in \mathbb{Q},$$

and then, if $\ell_1, ..., \ell_N$ are linearly independent over $\mathbb{Q}$, it follows $\alpha_i = 0, i = 1, ..., N,$. Thus, ℓ_j/ℓ_i, $j = 1, ..., N$, are linearly independent over $\mathbb{Q}$, too.

Combining these results with Corollary A.2 we obtain

Corollary A.10. *Let $\ell_1, ..., \ell_N$ be positive numbers. Then*

1) *If for all values $i, j = 1, ..., N$, $i \neq j$, the ratios ℓ_i/ℓ_j belong to $\mathbf{B}_\varepsilon$ then there exists a constant $C_\varepsilon > 0$ such that*

$$\mathbf{a}(\lambda) \geq \frac{C_\varepsilon}{\lambda^{N-1+\varepsilon}},$$

for every $\lambda > 0$. In particular, if all the ratios belong to $\mathcal{F}$, this inequality holds with $\varepsilon = 0$.

2) *If the numbers $\ell_1, ..., \ell_N$ satisfy the conditions (S), then, for each $\varepsilon > 0$ there exists a constant $C_\varepsilon > 0$ such that*

$$\mathbf{a}(\lambda) \geq \frac{C_\varepsilon}{\lambda^{1+\varepsilon}},$$

for every $\lambda > 0$.

Proposition A.11. *Let (ω_n) an unbounded sequence of positive solutions of the equation*

$$\sum_{i=0}^{N} \left(\cos \ell_i \omega \prod_{j \neq i} \sin \ell_j \omega \right) = 0 \tag{A.9}$$

and assume that the numbers $\ell_0, ..., \ell_N$ satisfy the conditions (S). Then, for every $\varepsilon > 0$ there exists a constant C_ε such that for every $n \in \mathbb{N}$ and every $i = 0, ..., N$, the following inequality is true

$$|\sin \ell_i \omega_n| \geq \frac{C_\varepsilon}{\omega_n^{1+\varepsilon}}.$$

Proof. It is similar to the proof of Proposition A.1. We will show that there exists a constant $C > 0$ such that, for every $i = 0, ..., N$,

$$\prod_{j \neq i} ||| \frac{\ell_j}{\ell_i} \mathbf{m}^i(\omega_n) ||| \leq C |\sin \omega_n \ell_i| . \tag{A.10}$$

This, in account of the Schmidt's theorem, give the assertion of the proposition.

In order to prove (A.10), it is sufficient to see that, if $|\sin \omega_n \ell_i| \to 0$ then

$$\prod_{j \neq i} ||| \frac{\ell_j}{\ell_i} \mathbf{m}^i(\omega_n) ||| \to 0. \tag{A.11}$$

Indeed, if $|\sin \omega_n \ell_0| \to 0$ (we have taken $i = 0$ to simplify the notations) then from the equality (A.9) it follows that

$$\prod_{j=1}^{N} |\sin \ell_j \omega_n| \to 0. \tag{A.12}$$

On the other hand, the inequality (A.5) obtained in the proof of Proposition A.1, allows us to ensure that

$$\prod_{j=1}^{N} ||| \frac{\ell_j}{\ell_0} \mathbf{m}^0(\omega_n) ||| \leq C |\sin \omega_n \ell_0| + \frac{1}{2^N} \prod_{j=1}^{N} |\sin \ell_j \omega_n| .$$

From this, based on (A.12) and the fact $|\sin \omega_n \ell_0| \to 0$, we obtain the convergence (A.11). This proves the assertion.

References

1. F. Alabau. Stabilisation frontière indirecte de systèmes faiblement couplés. *C. R. Acad. Sci. Paris, Série I*, 328:1015–1020, 1999.
2. F. Ali Mehmeti. A characterisation of generalized C^∞ notion on nets. *Int. Eq. and Operator Theory*, 9:753–766, 1986.
3. F. Ali Mehmeti. Regular solutions of transmission and interaction problems for wave equations. *Math. Meth. Appl. Sci.*, 11:665–685, 1989.
4. F. Ali Mehmeti. *Nonlinear waves in networks*. Mathematical Research, 80, Akademie-Verlag, Berlin, 1994.
5. F. Ali Mehmeti, J. von Below and S. Nicaise, (Eds). *Partial differential equations on multistructures*. Lecture Notes in Pure and Applied Mathematics, 219, Marcel Dekker, 2001.
6. F. Ali Mehmeti and V. Régnier. Splitting of energy and dispersive waves in a star-shaped network. *Z. Angew. Math. Mech.*, 83:105–118, 2003.
7. K. Ammari and M. Jellouli. Stabilization of star-shaped networks of strings. *Differential Integral Equations*, 17:1395–1410, 2004.
8. K. Appel and W. Haken. Every planar map is four colorable. *Illinois Journal of Mathematics*, 21:429–567, 1977.
9. S. A. Avdonin and S. A. Ivanov. *Families of Exponentials: The Method of Moments in Controllability Problems for Distributed Parameter Systems*. Cambridge Univ. Press, Cambridge, 1995.
10. S.A. Avdonin and W. Moran. *Simultaneous controllability of elastic systems*, pages 28–31. in: I.I. Eremin, I. Lasiecka, V.I. Maksimov, (Eds.), Proceedings of the International Conference on Distributed Systems, Ekaterinburg, Russia, 30 May–2 June, 2000. IMM UrO RAN, 2000.
11. S.A. Avdonin and W. Moran. Ingham type inequalities and Riesz bases of divided differences. *"Mathematical methods of optimization and control of large-scale systems" (Ekaterinburg, 2000), Int. J. Appl. Math. Comput. Sci.*, 11(4):803–820, 2001.
12. S.A. Avdonin and W. Moran. Simultaneous control problems for systems of elastic strings and beams. *Systems & Control Letters*, 44(2):147–155, 2001.
13. S. A. Avdonin and T. I. Seidman. Pointwise and internal controllability for the wave equation. *Appl. Math. Optim.*, 46:107–124, 2002.
14. S.A. Avdonin and M. Tucsnak. Simultaneous controllability in sharp time for two elastic strings. *ESAIM:COCV*, 6:259–273, 2001.

15. C. Baiocchi, V. Komornik, and P. Loreti. Ingham type theorems and applications to control theory. *Boll. Un. Mat. Ital.*, B 8 (II-B):33–63, 1999.

16. C. Baiocchi, V. Komornik, and P. Loreti. Généralisation d'un théorème de Beurling et application à la théorie du contrôle. *C. R. Acad. Sci. Paris, Série I*, 330:281–286, 2000.

17. C. Baiocchi, V. Komornik, and P. Loreti. Ingham-Beurling type theorems with weakened gap conditions. *Acta Math. Hungar.*, 97(1):55–95, 2002.

18. C. Bardos, G. Lebeau, and Rauch J. Sharp sufficient conditions for the observation, control and stabilization of waves from the boundary. *SIAM J. Control Optim.*, 30:1024–1065, 1992.

19. N. K. Bari. Biorthogonal systems and bases in Hilbert spaces. *Učen. Zap. Mosk. Gos. Univ.*, 4(148):68–107, 1951.

20. A. Beurling. *The Collected Works of Arne Beurling*, volume 2, Harmonic Analysis, L. Carleson, P. Malliavin, J. Neuberger, J. Wermer, (Eds.). Birkhäuser, Boston, 1989.

21. A. Beurling and P. Malliavin. On the closure of characters and zeros of entire functions. *Acta Math.*, 118:79–93, 1967.

22. U. Brauer and G. Leugering. On boundary observability estimates for semi-discretizations of a dynamic network of elastic strings. *Recent advances in control of PDEs. Control &Cybernetics*, 28:421–447, 1999.

23. H. Brézis. *Analyse fonctionnelle. Théorie et applications. 1983.* Collection Mathématiques Appliquées pour la Maîtrise, Masson, Paris, 1983.

24. N. Burq. Un théorème de contrôle pour une structure multidimensionnelle. *Comm. Partial Differential Equations*, 19:199–211, 1994.

25. N. Burq and G. Lebeau. Mesures de défaut de compacité, application su système de Lamé. *Annales Scientifiques de l' Ecole Normale Supérieure*, 34:817–870, 2001.

26. J.M. Cassels. *An introduction to diophantine approximation.* Cambridge Univ. Press, Cambridge, 1966.

27. C. Castro. Boundary controllability of the one-dimensional wave equation with rapidly oscillating density. *Asymptotic Analysis*, 20:317–350, 1999.

28. C. Castro and E. Zuazua. A hybrid system consisting of two flexible beams connected by a point mass: Well posedness in asymmetric spaces. In *"Elasticité, Viscoélasticité et Contrôle Optimal", ESAIM Proceedings, 2, 1997, 17-53.*

29. C. Castro and E. Zuazua. Une remarque sur les séries de Fourier non-harmoniques et son application à la contrôlabilité des cordes avec densité singulière. *C. R. Acad. Sci. Paris, Série I*, 332:365–370, 1996.

30. G. Chen, M. Delfour, A. Krall, and G. Payre. Modelling, stabilization and control of serially connected beams. *SIAM J. Control Opt.*, 25:526–546, 1987.

31. G. Chen and J. Zhou. The wave propagation method for the analysis of boundary stabilization in vibrating structures. *SIAM J. Appl. Math.*, 50:1254–1283, 1990.

32. R. Courant and D. Hilbert. *Methods of Mathematical Physics*, volume I. Interscience, New York, 1962.

33. R. Dáger. Observation and control of vibrations in tree-shaped networks of strings. *SIAM J. Control Optim.*, 43:590–623, 2004.

34. R. Dáger and E. Zuazua. Controllability of star-shaped networks of strings. *In: Bermúdez A. et al. (Eds), Fifth International Conference on Mathematical and Numerical Aspects of Wave Propagation, SIAM Proceedings*, pages 1006–1010, 2000.

35. R. Dáger and E. Zuazua. Controllability of star-shaped networks of strings. *C. R. Acad. Sci. Paris, Série I*, 332:621–626, 2001.

36. R. Dáger and E. Zuazua. Controllability of tree-shaped networks of vibrating strings. *C. R. Acad. Sci. Paris, Série I*, 332:1087–1092, 2001.

37. R. Dáger and E. Zuazua. Spectral boundary controllability of networks of strings. *C. R. Acad. Sci. Paris, Sèrie I*, 334(7):545–550, 2002.

38. B. D'Andréa-Novel and J. M. Coron. Stabilization of an overhead crane with a variable length flexible cable. *Special issue in memory of Jacques-Louis Lions. Comput. Appl. Math.*, 21:101–134, 2002.

39. B. Dekoninck and S. Nicaise. Control of networks of Euler-Bernoulli beams. *ESAIM:COCV*, 4:57–81, 1999.

40. B. Dekoninck and S. Nicaise. The eigenvalue problem for networks of beams. *Linear Algebra Appl.*, 314:165–189, 2000.

41. A. Doubova and E. Fernández-Cara. Some control results for simplified one-dimensional models of fluid-solid interaction. *Math. Models Methods Appl. Sciences*, 15(5):783–824, 2005.

42. C. Fabre and J. P. Puel. Pointwise controllability as limit of internal controllability for the wave equation in one space dimension. *Portugal. Math.*, 51:335–350, 1994.

43. H. Fattorini. On complete controllability of linear systems. *J. Diff. Eq.*, 3:391–402, 1967.

44. H. O. Fattorini. Estimates for sequences biorthogonal to certain complex exponentials and boundary control of the wave equation. *Lectures Notes in Control and Inform. Sci.*, 2:111–124, 1979.

45. H. O. Fattorini and D. L. Russell. Uniform bounds on biorthogonal functions for real exponentials and application to the control theory of parabolic equations. *Quart. Appl. Math.*, 32:45–69, 1974.

46. E. Fernández-Cara and E. Zuazua. On the null controllability of the one-dimensional heat equation with BV coefficients. *Computational and Applied Mathematics*, 21:167–190, 2002.

47. P. Grisvard. Contrôlabilité exacte des solutions de l'équation des ondes en présence de singularités. *J. Math. Pures Appl.*, 68:215–259, 1989.

48. M. Gugat, G. Leugering and E. J. P. G. Schmidt. Global controllability between steady supercritical flows in channel networks. *Math. Methods Appl. Sci.*, 27:781–802, 2004.

49. S. Hansen and E. Zuazua. Controllability and stabilization of strings with point masses. *SIAM J. Cont. Optim.*, 35:1357–1391, 1995.

50. A. Haraux and S. Jaffard. Pointwise and spectral control of plate vibrations. *Revista Matemática Iberoamericana*, 7:1:1–24, 1991.

51. L. F. Ho. Controllability and stabilizability of coupled strings with control applied at the coupled points. *SIAM J. Control Optim.*, 31:1416–1433, 1993.

52. V. C. L. Hutson and J. S. Pym. *Applications of Functional Analysis and Operator Theory*, volume 146 of *Mathematics in Science and Engineering*. Acad. Press, London, 1980.

53. J. A. Infante and E. Zuazua. Boundary observability for the space-discretizations of the one-dimensional wave equation. *Mathematical Modelling and Numerical Analysis*, 33:407–438, 1999.

54. A. E. Ingham. Some trigonometrical inequalities with applications in the theory of series. *Math. Z.*, 41:367–379, 1936.

55. E. Isaacson and H.B. Keller. *Analysis of Numerical Methods*. John Wiley & Sons, New York, 1966.

56. S. Jaffard, M. Tucsnak, and E. Zuazua. On a theorem of Ingham. *J. Fourier Anal. Appl.*, (3):577–582, 1997.

57. S. Jaffard, M. Tucsnak, and E. Zuazua. Singular internal stabilization of the wave equation. *J. Differential Equations*, 145:184–215, 1998.

58. A.G. Khapalov. Interior point control and observation for the wave equation. *Abstract and Applied Analysis*, 1(2):219–236, 1996.

59. V. Komornik. *Exact Controllability and Stabilization. The Multiplier Method*. Masson, Paris and John Wiley & Sons, Chichester, 1994.

60. V. Komornik and P. Loreti. Dirichlet series and simultaneous observability: two problems solved by the same approach. *Systems & Control Letters*, 48(3-4):221–227, 2003.

61. V. Komornik and M. Yamamoto. Upper and lower estimates in determining point sources in a wave equation. *Inverse Problems*, 18(2):319–329, 2002.

62. W. Krabs. *On Moment Theory and Controllability of One-Dimensional Vibrating Systems and Heating Processes*, volume 173 of *Lecture Notes in Control and Inform. Sci.* Springer-Verlag, Berlin, 1992.

63. J. E. Lagnese. *Recent progress and open problems in control of multi-link elastic structures.*, volume 209 of *Contemp. Math.*, pages 161–175. Amer. Math. Soc., Providence, RI, 1997.

64. J. E. Lagnese. Domain decomposition in exact controllability of second order hyperbolic systems on 1-d networks. Recent advances in control of PDE. *Control & Cybernetics*, 28(3):531–556, 1999.

65. J. E. Lagnese and G. Leugering. Dynamic domain decomposition in approximate and exact boundary control in problems of transmission for the wave equation. *SIAM J. Control Optim.*, 38(2):503–537, 2000.

66. J. E. Lagnese and G. Leugering. *Domain decomposition methods in optimal control of partial differential equations*. Birkhäuser Verlag, 148, Basel, 2004.

67. J. E. Lagnese, G. Leugering, and E.J. P.G. Schmidt. Control of planar networks of Timoshenko beams. *SIAM J. Control Optim.*, 31:780–811, 1993.

68. J. E. Lagnese, G. Leugering, and E.J.P.G. Schmidt. *Modelling, analysis and control of multi-link flexible structures*. Systems and Control: Foundations and Applications, Birkhäuser-Basel, 1994.

69. S. Lang. *Introduction to Diophantine Approximations*. Addison-Wesley, Reading, MA, 1966.

70. G. Lebeau. Contrôle de l'équation de Schrödinger. *J. Math. Pures Appl.*, 71:267–291, 1992.

71. G. Lebeau and E. Zuazua. Null controllability of a system of linear thermoelasticity. *Arch. Ration. Mech. Anal.*, 141:297–329, 1998.

72. G. Leugering. Reverberation analysis and control of networks of elastic strings. *Control of PDE and appl., Lect. Notes in Pure and Applied Math.*, pages 193–206, 1996.

73. G. Leugering and E. J. P. G. Schmidt. On the control of networks of vibrating strings and beams. In *Proc. Of the 28th IEEE Conference on Decision and Control, Vol. 3*, pages 2287–2290. IEEE, 1989.

74. G. Leugering and E.J.P.G. Schmidt. On the modelling and stabilization of flows in networks of open canals. *SIAM J. Control Optim.*, 41:164–180, 2002.

75. G. Leugering and E. Zuazua. On exact controllability of generic trees. In *"Control of Systems Governed by Partial Differential Equations"*, Nancy, France, March 1999, ESAIM Proceedings, vol. 8.

76. B. J. Levin. *Distribution of Zeros of Entire Functions*, volume 5 of *Translation of Math. Monographs*. AMS, 1950.

77. J.-L. Lions. Contrôlabilité exacte de systèmes distribués. *C. R. Acad. Sci. Paris, Série I*, 302:471–475, 1986.

78. J.-L. Lions. *Contrôlabilité Exacte Perturbations et Stabilisation de Systèmes Distribués*, volume I. Masson, Paris, 1988.

79. J. L. Lions. Exact controllability, stabilizability and perturbations for distributed systems. *SIAM Rev.*, 30:1–68, 1988.

80. J.-L. Lions and E. Magenes. *Problèmes aux limites non homogènes et applications*. Dunod, Paris, 1968.

81. A. López and E. Zuazua. Uniform null controllability for the one dimensional heat equation with rapidly oscillating periodic density. *Annales IHP. Analyse non linéaire*, 19:543–580, 2002.

82. G. Lumer. Connecting of local operators and evolution equations on networks. In *Lecture Notes in Math., 787, Springer Verlag, 1980, 219-234.*

83. G. Lumer. Connecting of local operators and evolution equations on networks. *C. R. Acad. Sc. Paris, Série A*, 291:627–630, 1980.

84. S. Micu and E. Zuazua. An introduction to the controllability of linear PDE. In *"Quelques questions de théorie du contrôle". Sari, T., ed., Collection Travaux en Cours, Hermann, pp. 69-157, 2005, to appear.*

85. S. Micu and E. Zuazua. Boundary controllability of a linear hybrid system arising in the control of noise. *SIAM J. Control Optim.*, 35(5):1614–1638, 1997.

86. L. Miller. Controllability cost of conservative systems: resolvent condition and transmutation. *J. Funct. Anal.*, 218(2):425–444, 2005.

87. L. Miller. Geometric bounds on the growth rate of null-controllability cost for the heat equation in small time. *J. Differential Equations*, 204:202–226, 2004.

88. L. Miller. How violent are fast controls for Schrödinger and plate vibrations? *Arch. Ration. Mech. Anal.*, 172:429–456, 2004.

89. M. Negreanu and E. Zuazua. Convergence of a multigrid method for the controllability of a 1-d wave equation. *C. R. Acad. Sci. Paris, Série I*, 338(4):413–418, 2004.

90. S. Nicaise. Spectre des réseaux topologiques finis. *Bull. Sc. Math, 2ᵉ Série*, 111:401–413, 1987.

91. S. Nicaise. *Polygonal interface problems.* Methoden und Verfahren Math. Physiks, 39, Peter Lang Verlag, 1993.

92. S. Nicaise. Boundary exact controllability of interface problems with singularities i: Addition of the coefficients of singularities. *SIAM J. Cont. Optim.*, 34:1512–1533, 1996.

93. S. Nicaise. Boundary exact controllability of interface problems with singularities ii: Addition of internal controls. *SIAM J. Cont. Optim.*, 35:585–603, 1997.

94. S. Nicaise and O. Penkin. Relationship between the lower frequency spectrum of plates and networks of beams. *Math. Methods Appl. Sci.*, 23:1389–1399, 2000.

95. S. Nicaise and O. Zair. Identifiability, stability and reconstruction results of point sources by boundary measurements in heteregeneous trees. *Rev. Mat. Complutense*, 16:1–28, 2003.

96. R. E. A. C. Paley and N. Wiener. *Fourier transform in the complex domain*, volume 19 of *Am. Math. Soc. Colloq. Publ.* Am. Math. Soc., New York, 1934.

97. O. M. Penkin, Yu. V. Pokornyi and E. N. Provotorova. On a vectorial boundary value problem. In *Boundary value problems, Interuniv. Collect. sci. Works, Perm' 1983, 64-70.*

98. K. D. Phung. Observability and control of Schrödinger equations. *SIAM J. Control Optim.*, 40:211–230, 2001.

99. J. P. Puel and E. Zuazua. Controllability of a multi-dimensional system of Schrödinger equations: Application to a system of plate and beam equations. In *"State and frequency domain approaches for infinite-dimensional systems"*. *Springer-Verlag, LNCIS 185, 1993, 500-511.*

100. J. P. Puel and E. Zuazua. Exact controllability for a model of multidimensional flexible structure. *Proc. Royal Soc. Edinburgh*, 123 A:323–344, 1993.

101. K. Ramdani, T. Takahashi, G. Tenenbaum and M. Tucsnak. A spectral approach for the exact observability of infinite dimensional systems with skew-adjoint generator. *Journal of Functional Analysis*, preprint.

102. R. M. Redheffer. Elementary remarks on completeness. *Duke Math. J.*, 35:103–116, 1968.

103. S. Rolewicz. On controllability of systems of strings. *Studia Math.*, 36(2):105–110, 1970.

104. K. F. Roth. Rational approximation to algebraic numbers. *Mathematika*, 2:1–20, 1955.

105. D. L. Russell. A unified boundary controllability theory for hyperbolic and parabolic partial differential equations. *Studies in Appl. Math.*, 52:189–221, 1973.

106. D. L. Russell. Controllability and stabilizability theory for linear partial differential equations. *SIAM Review*, 20:639–739, 1978.

107. D. L. Russell. The Dirichlet-Neumann boundary control problem associated with Maxwell's equations in a cylindrical region. *SIAM J. Control Optim.*, 24:199–229, 1986.

108. E. J. P. G. Schimdt. On the modelling and exact controllability of networks of vibrating strings. *SIAM J. Control. Opt.*, 30:229–245, 1992.

109. W.M. Schmidt. Simultaneous approximation to algebraic numbers by rationals. *Acta Math.*, 125:189–202, 1970.

110. M. Tucsnak and G. Weiss. Simultaneous exact controllability and some applications. *SIAM J. Control Optim.*, 38:1408–1427, 2000.

111. U. Ullrich. Divided differences and systems of nonharmonic Fourier series. *Proc. Amer. Math. Soc.*, 80:47–57, 1980.

112. J. von Below. A characteristic equation associated to an eigenvalue problem on c^2-networks. *Linear Alg. Appl.*, 71:309–325, 1985.

113. J. von Below. Classical solvability of linear parabolic equations on networks. *J. Diff. Eq.*, 72:316–337, 1988.

114. J. von Below. *Parabolic Network Equations*. Habilitation Thesis, Eberhard-Karls-Universität Tübingen, 1993.

115. N. Wiener. A class of gap theorems. *Ann. Scuola Norm. Sup. Pisa*, 3:367–372, 1934.

116. R.M. Young. *An Introduction to Nonharmonic Fourier Series*. Academic Press, New York, 1980.

117. X. Zhang and E. Zuazua. Exact controllability of the semi-linear wave equation. In *"Unsolved problems in mathematical systems and control theory"*, *Princeton University Press, 2004, pp. 173-178*.

118. X. Zhang and E. Zuazua. Polynomial decay and control for a 1-d model of fluid-structure interaction. *C. R. Acad. Sci. Paris, Série I*, 336:745–750, 2003.

119. E. Zuazua. Exact controllability for the semilinear wave equation in one space dimension. *Ann. IHP. Analyse non linéaire*, 10:109–129, 1993.

120. E. Zuazua. Controllability of the linear system of thermoelasticity. *J. Mathématiques pures et appl.*, 74:303–346, 1995.

121. E. Zuazua. Some problems and results on the controllability of partial differential equations. In *Progress in Mathematics, 169*, pages 276–311. Birkhaüser Verlag, 1998.

122. E. Zuazua. Boundary observability for the finite-difference space semi-discretizations of the 2-d wave equation in the square. *J. Math. Pures et Appliquées*, 78:523–563, 1999.

123. E. Zuazua. Controllability of partial differential equations and its semi-discrete approximations. *Discrete and Continuous Dynamical Systems*, 8:469–513, 2002.

124. E. Zuazua. Propagation, observation, and control of waves approximated by finite difference methods. *SIAM Review*, 47(2):197–243, 2005.

Index

Déjà parus dans la même collection

Déjà parus dans la même collection

Printing and Binding: Strauss GmbH, Mörlenbach